Energy and mineral resource systems:
an introduction

Energy and mineral resource systems

AN INTRODUCTION

B.A. TAPP
Senior Lecturer

J.R. WATKINS
Former Head

Department of Geology and Geological Engineering
Royal Melbourne Institute of Technology

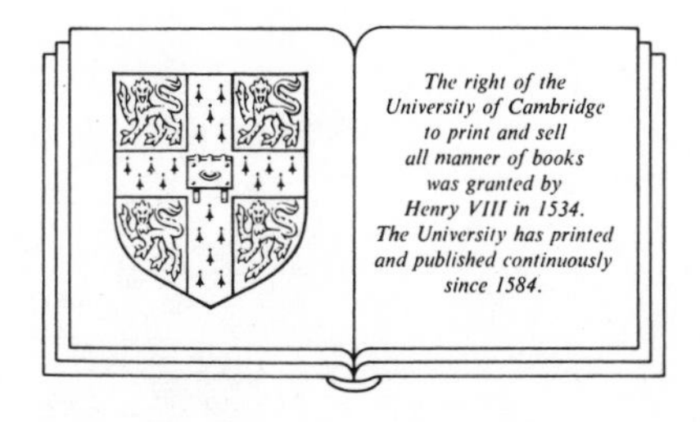
The right of the
University of Cambridge
to print and sell
all manner of books
was granted by
Henry VIII in 1534.
The University has printed
and published continuously
since 1584.

CAMBRIDGE UNIVERSITY PRESS
Cambridge
New York Port Chester
Melbourne Sydney

333.79
T

Published by the Press Syndicate of the University of Cambridge
The Pitt Building, Trumpington Street, Cambridge CB2 1RP
40 West 20th Street, New York, NY 10011, USA
10 Stamford Road, Oakleigh, Melbourne 3166, Australia

© Cambridge University Press 1990

First published 1990

Printed in Great Britain at the University Press, Cambridge

British Library cataloguing in publication data

Tapp, B.A.
Energy and mineral resource systems: an introduction.
1. Energy budget (Geophysics)
I. Title II. Watkins, J.R.
553 QC809,E6

Library of Congress cataloguing in publication data

Tapp, B. A.
Energy and mineral resource systems.

Bibliography
1. Energy minerals. 2. Mines and mineral resources.
I. Watkins, J. R. II. Title.
TN263.5.T36 1989 333.79 86-11722

ISBN 0 521 30287 0 hard covers

ISBN 0 521 31616 2 paperback

18 NOV 1991

MU

Contents

Preface

Educational systems, at least in the Western world, foster the idea that Nature is divisible into distinct facets or disciplines. For the purposes of study 'we' arbitrarily sub-divide the world in which we live into units – physics, chemistry, and such like. There is no doubt that these sub-divisions allow for the acquisition of information or data and allow theories to be developed and formulated. These sub-divisions are also constricting, however, in that a single discipline approach to a study can give a bias and hence distort our understanding of reality. It is only in relatively recent times that 'systems' have been studied as trans-disciplinary entities, and even here our knowledge is distorted since no earthly system is truly isolated. A 'systems approach' requires a philosophy that considers both the elements and their inter-relatedness in an operational framework. It is essentially holistic. This book attempts an overview of but two of the systems within Nature – the energy and mineral resource systems.

The boundaries of the system are in reality diffuse, being observer dependent and definable by a variable set of criteria. Even so, energy and mineral resources are foundational to civilisation and are both international and interdisciplinary in context. The systems comprise interactions from science, economics, sociology and politics, and for the global well-being of mankind it behoves us to at least attempt to comprehend them.

We do not pretend that this is a book of answers. It is a book, however, that we hope both stimulates and shows the diverse character of energy and mineral resource systems.

We have drawn material from a wide variety of sources and attempted to show how the systems can be developed and organised according to particular individual requirements. We are therefore indebted to many people for ideas, inspiration and constructive criticism. Two people, however, stand out in particular.

The first is Jonathan West who introduced us to this particular area of study and upon whose work we have drawn. In particular Chapters 2 and 3 quote liberally from his work and we express our indebtedness and gratitude for that permission. The second is the wife of one of us (B. T.) who not only tolerated this with rare good humour but also typed the original manuscript.

Our thanks should also be extended to Mrs Irene Pizzie who spent many hours patiently turning a draft into a book.

BT/JRW. Melbourne

Introduction

In a simplistic way a system can be conceived as a set of operations being acted upon and themselves acting upon one or more inputs. This process of throughput gives rise to the production of an output or outputs. By definition a system is therefore concerned with a flow, which may take the form of mass, energy or information. The physical world as such can be viewed as a network of intermeshed systems; this book attempts to analyse two closely inter-related systems: the energy system and the mineral resource system.

The analysis of any system involves the problems of boundary definition, the form and type of input variables and the process(es) operating within the system. Any definition and analysis of a system is to a large extent observer dependent. The content of the book therefore reflects the bias of the authors, and the intent of the book is not directed toward rigorous analysis of a variety of system options but to discuss the major themes inherent within the energy and mineral resource systems. The book is directed toward a wide ranging spectrum of professional and interested people, and it is hoped that it will give insight into two complex systems upon which society places great reliance. As such the book is divided into three parts; the first section concerns energy, the second mineral resources, and the third the methods of resource evaluation within these systems.

The boundaries to the energy and mineral resource systems and their relationship to other systems in an industrialised economy are shown in Figure A. The geosystem here defined solely as a source of natural geological materials is the determinant of energy and mineral reserves and resources – the distinction between reserves and resources will be clarified in later chapters. It could be argued that energy resources themselves are mineral resources. In one sense this is true since energy throughputs are initially defined from mineral resources such as coal. Energy is, however, a facilitator enabling society to pursue particular goals and objectives.

The geosystem is subject to what we term 'human directive'. The directive and the form that it takes are determined by the prevailing political, social and economic parameters prevailing within any one particular society. This determines the use, and rate of usage, of the resources available to mankind. Systems can therefore be said to have motives and the problem is one of control. Some societies advocate some form of 'free market' control, others favour totalitarian state control, whereas a few advocate no control.

Many societies tend to regard many materials as both energy and mineral resources. In the Australian economy, for example, coal is regarded as both a mineral commodity and an energy source. It is regarded as a mineral commodity when it is mined

Figure A. Systems interaction in an industrialised complex.

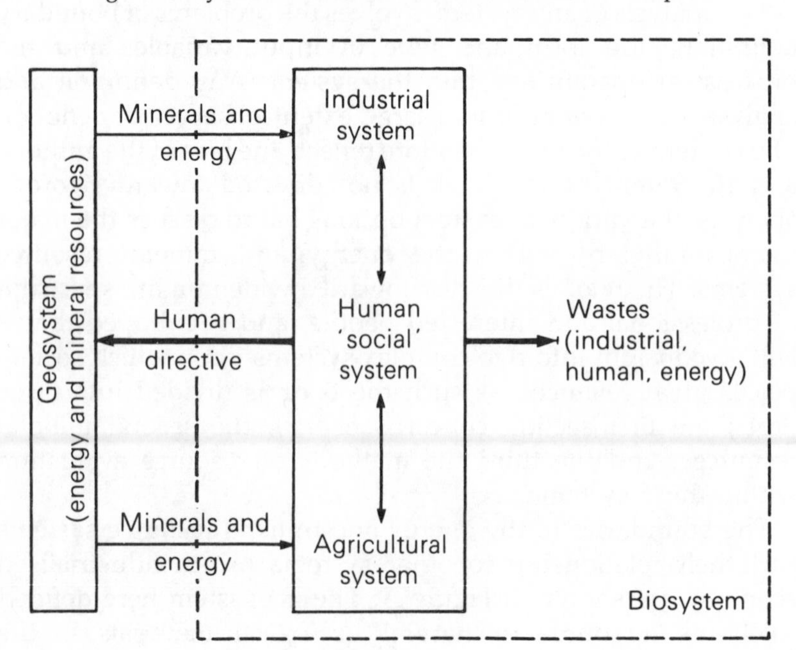

and exported with the sole purpose being the production or obtaining of foreign revenue. It is also regarded as an energy source when utilised domestically as a source of fuel either in terms of electricity generation or direct industrial feedstock.

The industrial complex shown in Figure A demonstrates a particular problem of systems interfacing. The figure shows that the generalised industrial complex is essentially a closed loop system of material transformation. Industrial nations are nothing more than simply material processing machines. The economic problems within the complex are generated when mutually interacting systems operate at different rates. These types of problems exist at both national and international level thereby generating destabilising forces threatening both levels of economies. Differing speeds of operation of the industrial complexes between countries, or states within a country, tend to exacerbate social and economic differences.

In addition, the operational differences existing between the industrial and agricultural systems cause considerable imbalance in terms of ecological stability. This inevitably flows on to the economic and social systems.

The purpose of this book is to analyse two systems within the industrialised complex and to outline the problems of development and growth. Two further points need clarification before this can be done; the first is a definition of the 'nature' of energy and mineral resource systems, and the second is the levels of analysis that are definable.

The global and national energy and mineral resource systems are not composed of single industrial units but represent a complex aggregate of many industries such as oil, gas, coal, iron ore, aluminium, etc. Each of these possesses its own markets and conditions of supply and demand. These parameters can, and do, vary to a very large extent between industries. The nature of these systems is such that nationally, production is obtained from a small number of large units and only a small percentage of the community is engaged in the production–marketing process.

There also needs to be caution in attempting to extrapolate the performance of any one of the industries in the systems on the basis of past performance. For example, the past few decades have witnessed the change from many small producing units to large companies and trans-national consortia. The minerals industry in particular is cyclic in character experiencing periods of 'boom' and 'bust' conditions. Changing economic conditions, the discovery of new reserves, resources or substitutions, and the introduction of new technologies all serve to cause major fluctuations in the levels of activity.

The second point, the 'levels' of analysis, relates to the varying perceptions of energy and mineral resource systems. The simplest perception occurs at the individual or small group level. Figure B details the energy discourse at this level. The main concerns are limited to the acquisition of resources to maintain a small personal life support system related to work, home and recreation. This is the 'I' network. 'Them' is the group representing the major and/or controlling interests. The barrier consists of government(s) and the media; the former being perceived as the provider mechanism of resources and the latter, the communication linkage.

It is at the government level that the second perceptual stage is defined. In Australia the government policy themes have been those of provision. To this end emphasis has been upon the acquisition of supply sources and maintenance of supply continuity. The processes of government and the economic calculus utilised does not conform to any resource system; a fragmented sectarian approach therefore results as is shown in Figure C. The problem in Australia is compounded by the fact that each state has its own government with a duplicate set of

Figure B. The energy discourse at a 'local' level. Adapted from Arnoux (1981).

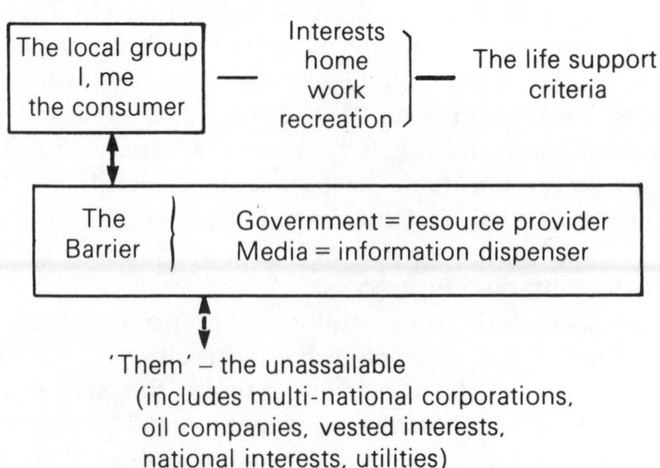

departments and committees. At the federal level in Australia seven ministerial departments are concerned with the formulation of energy policy and implementation. These ministries are advised by 30 different advisory and statutory bodies. Although there is some cross-fertilisation in terms of personnel membership similarity in various committees, each of these bodies has a sectarian and biased perspective.

The third and final level of analysis is the 'expert' definition of a conceptual model analysing the major forces shaping the character of society. These conceptual models take the form of one-, two- or three-dimensional flow charts of the systems of cause and effect pathways which determine the main features of society. These models can be defined in terms of the energy system only and its effects on the economic calculus (Musgrove *et al.*, 1983; Brain & Schuyers, 1981).

Other models are far more embracing attempting to define a complete global causal system network (Watt, 1979). Watt's model operates at three levels: the local city or state level, the national level and the international level. These hierarchical levels form one of the dimensions in the analysis. Another dimension is defined by a causal pathway which includes

Figure C. Energy policy and Australian Federal Government, 1987.
Source: Dept. Annual Reports.

Energy policy: formulation and implementation ←

Ministries	Statutory bodies	Advisory bodies
Treasury	1	–
Primary industry and energy	4	16
Environment	–	2
Transport	–	10
Industry and commerce	2	–
Business and consumer affairs	1	2
Science and technology	2	–

ultimate causes, key driving factors, key driving processes, immediate and ultimate effects. The third dimension or axis differentiates between seven differing categories of ultimate causes – the physical environment, the non-renewable resources, renewable resources, human population characteristics, economic factors and parameters, politico–governmental factors and the socio–cultural–psychological parameters.

Each of these levels, or ways of thinking about energy and mineral resource systems, reflects a differing perception of these complex systems. Each is useful in particular contexts for facilitating communication about system behaviour. They also provide the basis for formulating policies and programmes for the development of or in response to forecast scenarios. The usage of computer programmes allows simulations to be made of the consequences of particular policy formats.

A scenario is a schema or device for ordering the perceptions about alternative environments within which policies are implemented. Scenarios are not predictions or variations of a base case and neither are they generalised views of particular desired or feared futures. A scenario can be one of three things. It can be a fundamental assessment of current trends or ideologies and an expression of plausible outcomes. With respect to the energy and mineral resource systems it can also include consistent and complementary expressions of plausible futures including parameters such as the socio-economic, political and technological environments. Lastly, it can be a custom designed and focused description of the plausible outcome of a desired alternative environment (West, 1978).

Part 1 defining the energy system is split into five chapters. The first defines the elements of the energy system; the second the historical perspective; the third provides an outline of current prognoses; the fourth defines the choices available; and the fifth describes some of the social factors involved in the scenarios.

Part 2 defining the mineral resource system is split into three chapters. Chapter 6, the first chapter in this part, defines the elements of the system; Chapter 7 defines the historical and current perspective; and Chapter 8 details the prognoses and the scenarios.

Part 3 defines the principles of resource evaluation. Chapter 9 describes the basics of exploration technologies and briefly mentions the sequences of exploration techniques. Integration is the theme or keyword in future exploration technology. The last chapter, the tenth, outlines some of the major themes that will have to be addressed for the definition of future scenarios in energy and mineral resource systems.

Part 1 The energy resource system

1

Energy in the human environment

In his book *What Men Live By* Tolstoy posed three basic questions: what is given to man?; what is not given to man?; and lastly, what does man live by? Since 1973 world economies have come to recognise the pivotal role that energy plays in our societal structures, and answers can be given to these three questions with regard to energy. Mankind has been given a wide variety of energy sources; mankind does not have vast amounts of cheap, available low-entropy energy; and finally mankind lives by consuming energy (Finney, 1976). The energy of a system is definable as the ability of that system to do or perform work. The energy content of a system is in essence the amount of work stored in the system. Power (performance) is definable as the amount of work done in unit time: the unit of power is the watt. Our theme is therefore the energy flow or capacity for production within a nation or social unit.

However, it was not until the actions of the Organisation of Petroleum Exporting Countries (OPEC) in 1973 that the role of energy in global economies was brought into focus. The series of then sharp increases in world oil prices brought to prominence a sudden, widespread recognition of the finiteness of global liquid fuel supplies. Apart from the new and unexpected pressures that were imposed on national economies by the rising oil prices, governments were actively stimulated into giving serious consideration to the very real problems of securing future energy supplies. As a result the term 'energy policy' came into vogue and quickly became a preoccupation of governments at all levels and also various institutional and industrial bodies. It has now come to be recognised that the

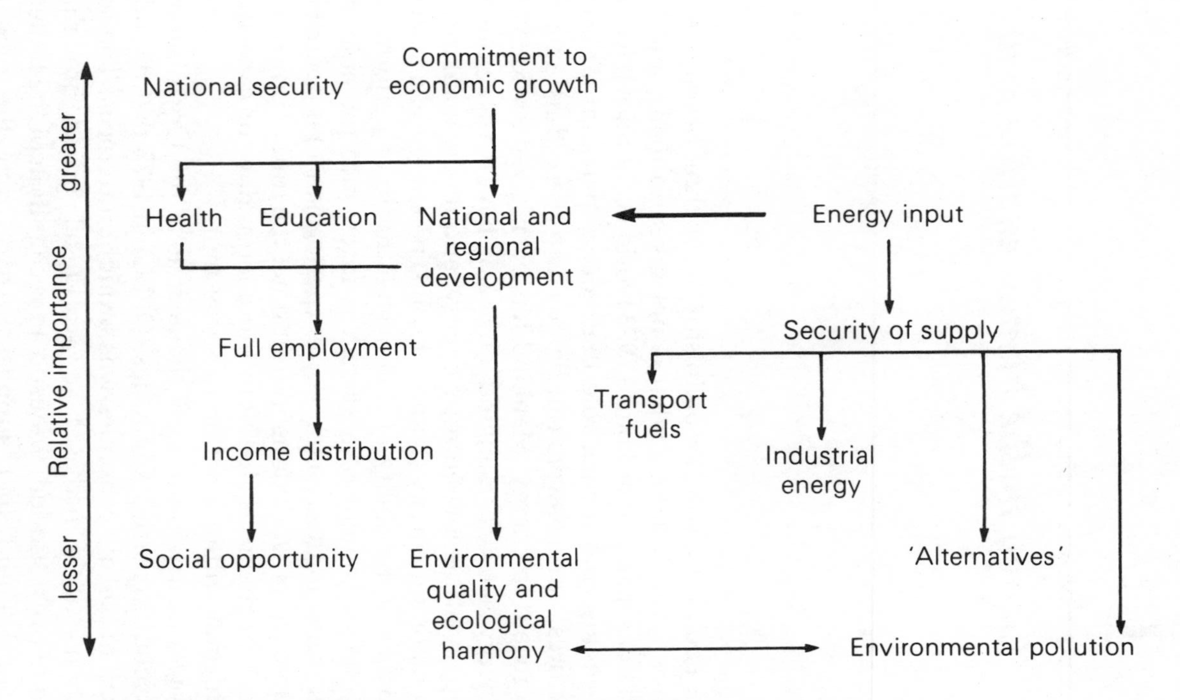

Figure 1.1. The relationship between national and energy objectives. Adapted from O'Riordan (1976).

level of 'economic development' achievable by a community is ultimately constrained by the availability of energy and in particular low-entropy energy.

Despite this neither energy management nor environmental planning ranks amongst the major goals of modern industrialised societies. Energy is an essential prerequisite to our technological societies, but so also is the sustainability of our ecosystems. The key question really is, how can a reasonable level of energy usage, at a properly defined cost such as replacement value, be maintained without causing permanent environmental degradation? Environmental protection is essential to our social and economic well-being, but the perennial problem facing policy makers and decision makers is that many of the environmental impacts of energy technologies are neither known nor readily quantified (Ophuls, 1977).

It is not unreasonable to assume that the spread of industrialisation and modern technologies to the 'less developed' regions of the world will greatly increase energy demand. The only problems to be solved in this context are the rates and ultimate extent of energy demand growth patterns. The actual make-up of energy resources required and available will be subject to economic and political constraints. The actual choices made 'today' as to the particular energy systems developed for the future will greatly affect the environmental conditions for decades to come. But at the same time the 'environmental' factors are major considerations in formulating energy systems. But as Collingridge (1980) has endeavoured to show, there is a disparity between our level of technical competence and our understanding of how the fruits of this competence affect human society.

The energy policies of many industrialised nations now tend to reflect considerations such as:

- the levels of economic activity and population trends;
- the relative costs of energy resources, including considerations of various technological developments;
- the environmental and conservational parameters.

The interaction between energy and national goals is shown in Figure 1.1.

The word 'energy' has many different meanings, and although it is true to say that modern society is facing a period of potential energy shortage, the shortage is of a particularly subtle kind. For example, there are advocates of energy conservation schema, yet 'energy is always conserved'. What is meant is that societies require 'technically useful energy' for which the term 'fuel' is perfectly adequate.

The fuels available to man are quite large; a qualitative and semi-quantitative indication of these resources is given in Table 1.1. Low-entropy energy sources such as oil have economic importance as fuels, yet they are statistically unlikely states of matter and so economic value is, in reality, an expression or representation of the scarcity (Finney, 1976).

The close relationship that exists between energy usage in production systems and the economic activity these systems support facilitates the interpretation of the changing patterns of the development of human civilisations. This can be done by defining the changing patterns of energy usage. The daily consumption of energy *per capita* for various 'societial types' is given in Figure 1.2. This figure is both a space and time reflection of human development: it reflects both the current geographic distribution of societal types and also the development phases of Western-style industrial societies.

With the advent of an agricultural revolution and mastery of solar-energy capture and grain storage, primitive or

Table 1.1 *Energy sources available to man*

Source	Amount (J)	
Deuterium in the oceans	10^{30}	Increasing entropy ↓
Uranium and thorium	10^{26}	
Coal, oil and gas	10^{24}	
Solar energy (Absorbed sunlight)	10^{24}	
Oceanic tides	10^{18}	
Wind power	10^{18}	
Hydro-electric power	10^{18}	
Energy from plants, wood	10^{16}	

Adapted from Hoyle (1977).

Note:
(i) Deuterium technologies are not yet available
(ii) Only a small fraction of incident solar, wind and tidal energy is harnassable
(iii) For coal, oil, gas, final amounts will depend on future exploration success

palaeotechnic man then becomes the advanced or neotechnic, agricultural man. A further industrial revolution, such as characterised and transformed western civilisation in the 18th and 19th centuries, defines industrial man. Further technological development and impetus, such as is current in the USA, then defines technological man. The effect of this trend toward greater industrialisation, which currently is the goal of the majority of nations, can be seen in a comparison of the relative energy consumption for each of the societal types.

Assuming palaeotechnic man has an energy consumption of one unit then the comparative consumptions are: palaeotechnic, 1; neotechnic, c3; industrial, c8; technological, c22. The progression, especially from 'industrial' to 'technological' man has given rise to the recent acceleration in the exploitation of the world's non-renewable resources. Human labour has been increasingly replaced by high-energy-cost machines. New high-energy-cost materials, such as aluminium and plastics, have replaced those with a lower energy cost such as steel and wood. Increased personal comfort, mobility and a high level of waste, have contributed to the large rise in energy consumption *per capita*.

At the present time this high level of energy usage is reliant upon limited stocks of non-renewable low-entropy fossil fuels. The realisation of the finite nature of fuel stocks has presented all governments with the problem of defining variable usage patterns and/or finding alternative fuel sources.

Global correlations between national wealth and consumption of low-entropy energy are well-known and have been widely publicised and criticised. The normal method of displaying this correlation is by means of a scattergram plotting energy consumption *per capita* against either the gross national product (GNP) or the gross domestic product (GDP). This methodology is not strictly applicable since in most cases the data used to generate the global energy regressions are highly selective.

Further criticisms may be made concerning the actual nature of the data. Data on total energy consumption are difficult to obtain since the majority of available energy data exclude such items as domestic, or non-commercial consumption. This tends to undervalue the consumption data for third world and developing nations, since the ratio of commercial to non-commercial energy consumption tends to be small. Also,

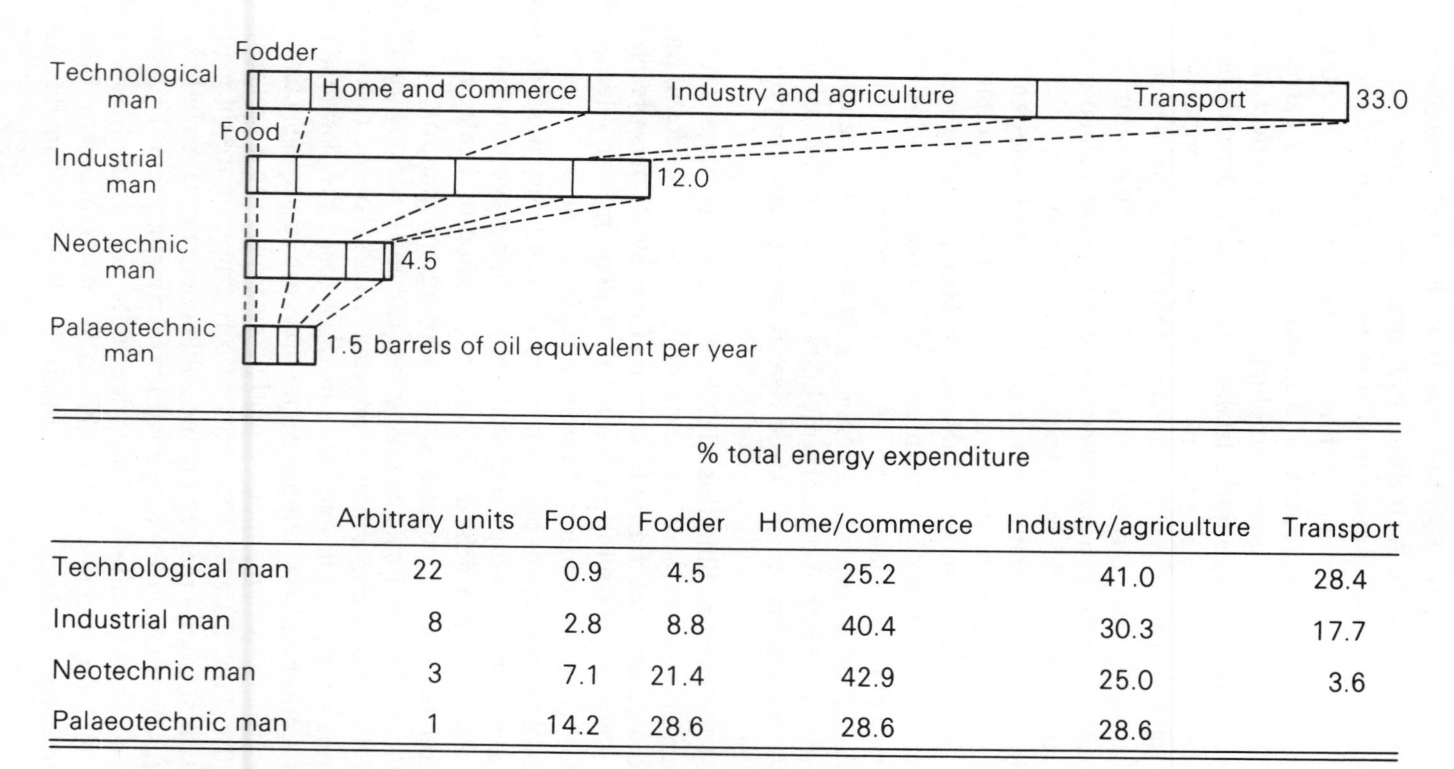

		% total energy expenditure				
	Arbitrary units	Food	Fodder	Home/commerce	Industry/agriculture	Transport
Technological man	22	0.9	4.5	25.2	41.0	28.4
Industrial man	8	2.8	8.8	40.4	30.3	17.7
Neotechnic man	3	7.1	21.4	42.9	25.0	3.6
Palaeotechnic man	1	14.2	28.6	28.6	28.6	

Figure 1.2. Yearly energy consumption *per capita*. Source: Messel (1979).

conversions used in the United Nations statistics generally ignore fuel-quality variants, and the effects of differing end-use efficiencies. In many cases, the appropriate conversion factors relating to the generation of primary electricity are either not available or not used (Finney, 1976).

The equivalence of the GDP and GNP indices is also a matter of some dispute. These indices are defined in terms of exchange rates which are based on internationally traded commodities. Again this mitigates against the third world and developing countries, which, with a predominantly internal agricultural society, may have little 'voice' in international trade. Political differences of style of government will also affect the index definition. There are, however, basic links between energy, the economic system and the environment. In many ways energy has created a cultural norm in society and imbalances between the living standards of the developing, developed and impoverished regions.

Technological development and industrialisation has become extremely energy intensive. About 99% of the driving force in modern industry is geochemically stored and transformed energy. Since 'market forces' fix the relativity of costs regarding human and natural energy, fossil fuels have been substituted wherever possible for human labour.

Unfortunately many governments have reasoned for a long time that a 'healthy' economy and a strong technological base are requisite for economic and social well-being. This is only true when unlimited supplies of resources are freely available. In the end analysis, however, it is not economics that matters but resource availability; for example, energy resources make other resources available.

The energy discourse is very complex, and the interwoven patterns of energy usage in our society show that the 'progression' of civilisation appears to be associated with a vicious spiral of ever-lengthening chains of dependency and an ever-increasing degree of technological addiction. As Figure 1.3 shows the flow of energy and materials through an economic system generates a large amount of waste. In fact, in 'industrial' nations there has now emerged the 'energy industry', an example of which is given in Figure 1.4.

In a modern industrial economy such as currently exists in Australia, the production of goods and services is a highly

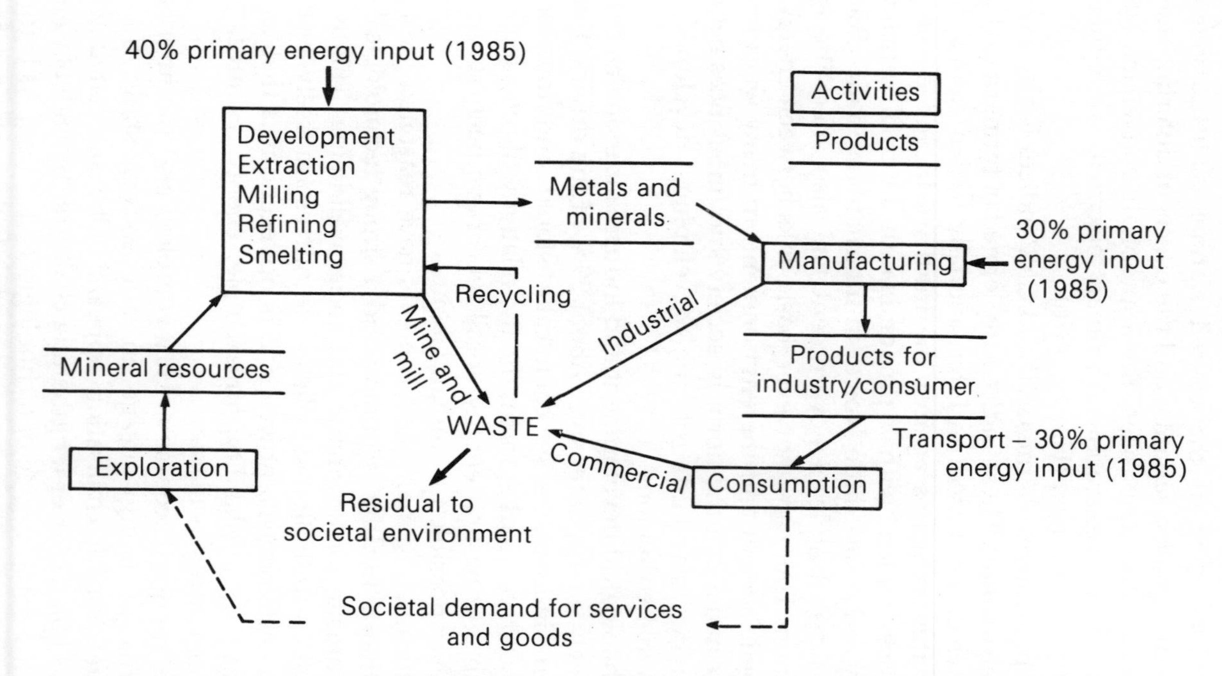

Figure 1.3. The flow of energy and materials through the economic system in Australia.

complex network of processes converting raw materials into final usable products. Resources flow through this network from process to process and machine to machine until they are finally converted to marketable goods for both domestic and overseas consumption. One of the resources entering this network, and one which is vital to the entire fabric of the economy, is energy. Part of the energy is processed by the network into an energy form purchased by the final buyer, whilst the remainder is used within the network to contribute to the actual production process for the whole range of goods and services (Becker, 1980).

The past decade has seen the emergence of two key factors which have contributed much to the creation of a degree of instability for this network. The first relates to the conditions governing the supply of energy, and in particular the problems surrounding continued supplies of oil, which threatens to disrupt the entire fabric. The second is definable in terms of the social changes which are reflected in the demands made upon the system and the methodology of its management.

Energy is but one resource, albeit a very vital one, which enters the complex production network. A small part of this energy input is processed directly into final consumer products

Figure 1.4. The energy industry. Adapted from Rawlings *et al.* (1983).

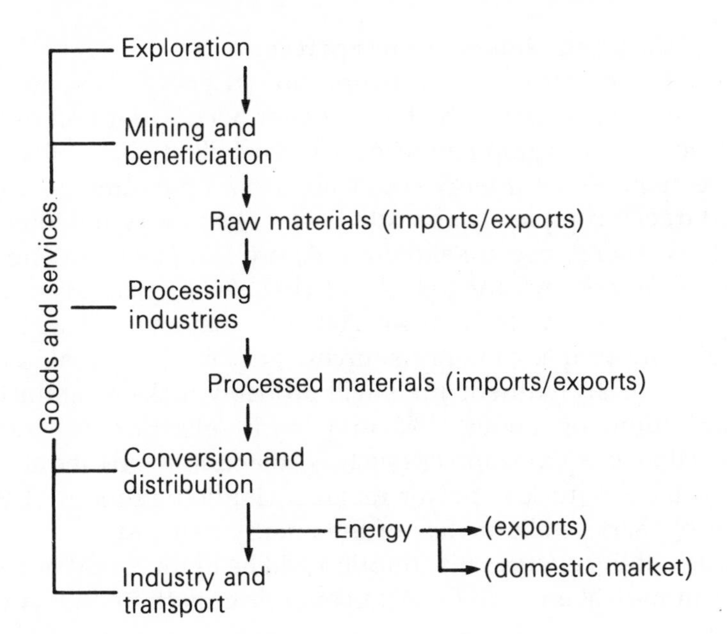

– coal to electricity and crude oil to motor spirit. By far the greater part, however, is used within the network to power the individual processes and to provide transport for materials and components from one locale to the next. The energy input and throughput is an integral part of any economic system and especially so in the mineral and mineral processing industries. The commitment to 'growth' scenarios will 'see' substantial demands placed upon energy resources and the pace of the changes in social and economic systems will be quite rapid (Becker, 1980).

The reasons for this are quite readily outlined. In the first place energy has ceased to be considered in purely technical terms. No longer is it regarded merely as a resource input, but is considered in the terms of broader economic, social and political implications. The rate of change in the industrial and energy environment has now of course increased quite substantially and this imposes great stress upon the management structure of the energy throughput. Lastly, and perhaps most importantly, the future supply of energy is globally uncertain – a condition that did not exist a decade ago. As Becker (1980, p.10) asks, 'has the interdependence between energy and society moved from energy requirement being a technical resultant of society's basic needs into the reverse?'

1.1 Uses and abuses of energy resources

The current world consumption of energy is some 340 EJ *per annum*. To attain the level of energy consumption of the USA for a global population of say ten billion some 50 years hence requires an energy input about 17 times this amount. This target is completely non-attainable. The history and forecast of global energy use are shown in Figure 1.5. The main energy sources at present are: oil, 45%; coal, 26%; gas, 18%; and nuclear, 0.5%. By the year AD 2000 oil, coal and gas will still remain the primary energy sources, *viz.* oil, 37%; coal, 24%; and gas, 16%. Synthetic fuels will probably make a significant contribution of about 4% and hydro-electric and other renewable energy resources (such as solar and wind) about 9%. The future of nuclear power remains clouded although it will probably service about 10% of the energy demand.

Much of the energy consumption will be in the world outside the communist areas (WOCA). Economic growth in WOCA has been fuelled since the Second World War by cheap and abundant energy, primarily oil. In 1973–4 came the first oil shock when OPEC signalled the end of cheap oil. This 'shock' raised the oil price from around US $3 per barrel to about US $10 per barrel. The onset of these steep price rises was due to enormous increases in demand and the political crises of the Middle East oil producers. In 1979 the second oil 'shock' was brought on due to the disruption of Iranian oil supplies. This second 'shock' took the price of oil to US $31 per barrel in 1981 (Mackrell, 1983).

These two periods of oil supply instability inevitably led to a follow-on fall in the demand for oil. This fall in demand was brought on indirectly by a downturn in the global economy and directly through a variety of measures such as enforced conservation, increased efficiency and fuel substitution. One of the major consequences of these measures has been the change in energy intensity (the ratio of energy demand to GDP). As Figure 1.6 shows, the initial oil price rises and disruptions of the early 1970s caused a very sharp fall in both energy and oil

Figure 1.5. World energy usage and sources.

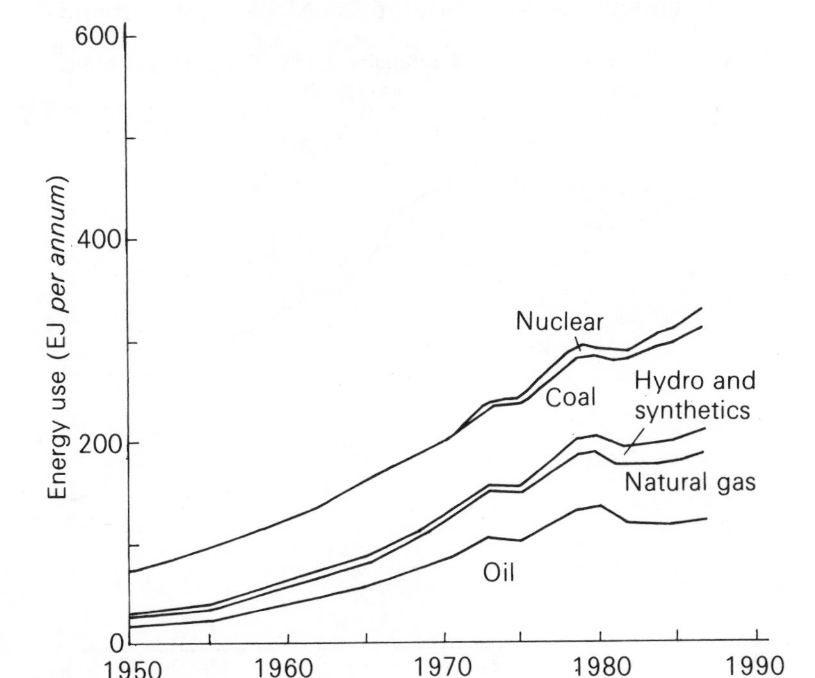

intensities. Prior to this time the energy intensity had been about unity.

The forces that have changed the energy intensity in WOCA have been of two types; an essentially short-term reversible series of factors, and a longer-term virtually permanent series of factors. These are outlined in Table 1.2. Perhaps the factor with the longest response time but the greatest potential is that relating to technological improvements. The lead time requirements for the implementation of new technologies can be quite substantial. As Mackrell (1983, p.80) states: 'The overall efficiency of energy using equipment has improved by 10% since 1973 and is likely to continue at better than 1% per annum. This could mean another 20–30% improvement by AD 2000 – equivalent to a further oil saving of 10–15 Mb/d'. (Mb/d = million barrels per day).

Currently much uncertainty exists over government responses to lower oil prices, especially in the short term. Again quoting Mackrell (1983, p.80):

> 'For oil, the second oil shock caused a drop in demand over the period 1979 to 1982 of 6.3 Mb/d instead of an increase of some 2.2 Mb/d had the 1979 oil intensity been maintained. Of this total of 8.5 Mb/d "lost demand",

Figure 1.6. Energy and oil intensity in WOCA 1960–81. Source: Mackrell (1983), *BP Statistical Review* (1987).

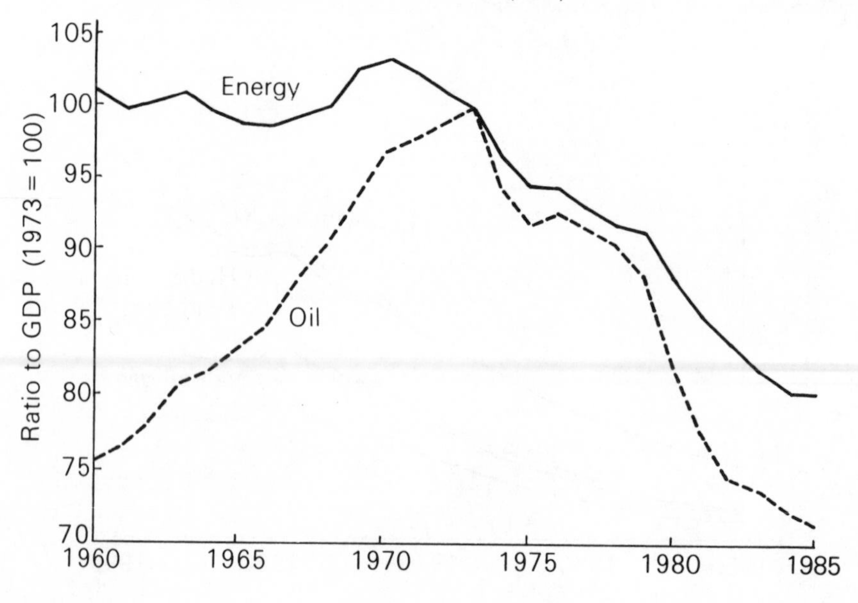

> some 5 Mb/d is attributable to Reversible factors, but the Permanent factors including substitution may be expected to continue to shade oil demand by some 2–3% *per annum*. Allowing for some interaction between factors, the long-term potential for reduction in oil demand could be some 15–30 Mb/d off what consumer demand in AD 2000 might have been without these "conservation" measures.'

The reduced demand for oil after 1979, combined with the recent decline in oil prices in 1983 and 1984, have dramatically changed perceptions of future prospects for alternative energy scenarios including what could be termed 'non-conventional oil'. Despite this the world is still basically oil hungry. From the foregoing it is easily discerned that the world will soon be populated by more people than the conventional energy sources can sustain, and this will be true whether or not the 'developing nations' achieve the current standards of material well-being of the 'developed nations'. The geographic distribution of energy consumption highlights the large discrepancies in energy consumption between nations of the varying economic determinants, i.e. developed and developing. What commonly exacerbates the problem is the tendency of developed nations to participate in what Collingridge (1980) defined as the 'hedging circle' when it comes to energy planning. The problem is simply that initially it is cheaper to increase the energy supply rather than attempt to increase usage efficiencies. To allow for supposed future growth, supply is increased and the scenario is effectively self-fulfilling as demand increases to take up the supply. The hedge, in reality, means that the energy system

Table 1.2 *Factors involved in energy intensity changes*
Source: Mackrell (1983)

Short-term/reversible factors	Long-term/irreversible factors
Immediate consumer response (the utilisation of alternatives, or doing without)	Industrial restructuring (reducing energy intensiveness of industry processes)
Stocking and destocking	Technical improvements
Changing fuel priorities and fuel mix sequences	Permanent movement of industry to 'cheap' sources of energy
Temporary substitution	Permanent substitution

loses flexibility and becomes increasingly difficult to control. Figure 1.7 defines the process, which constitutes one of the greatest abuses of energy resources. The morality of growth constitutes another.

1.2 Energy needs and greeds

Global economies are imprisoned behind the triad of growth, progress and development. The onward march toward these eternal myths is measured in terms of various indices, technology itself being the economic calculus. Money, or its equivalent, actually allows the calculation of the numbers, such as GDP, GNP, CPI, etc., but there is an underlying, oft forgotten concept, the idea of the perfectibility of mankind and society. There is a striving toward perfection, an ever-upward and onward trend. The whole unwieldy edifice runs on energy; the greater the input of energy, the greater the illusion of progress.

As Arnoux (1979) describes it, modern life consists of the myth of consumption and the consumption of a myth. Now according to Toynbee (quoted in Capra 1982, p.7) the 'genesis of a civilisation consists of the transition from a static condition to dynamic activity'. This change is brought about by a process of challenge and response, the challenge itself arising internally or via the influence of adjacent civilisations. Toynbee continues: 'The civilisation continues to grow when its successful response to the initial challenge generates cultural momentum that carries

Figure 1.7. The hedging circle in energy planning. Source: Collingridge (1980).

Error cost of increasing energy usage efficiencies greater than error cost of increasing energy supply
Hedge: increase supply to allow for GDP growth
Additional demand scenario is self-fulfilling since consumers adjust to cheap plentiful energy supplies
Control of energy system becomes low priority. Demand forecasts are inflated and supply is liberally planned

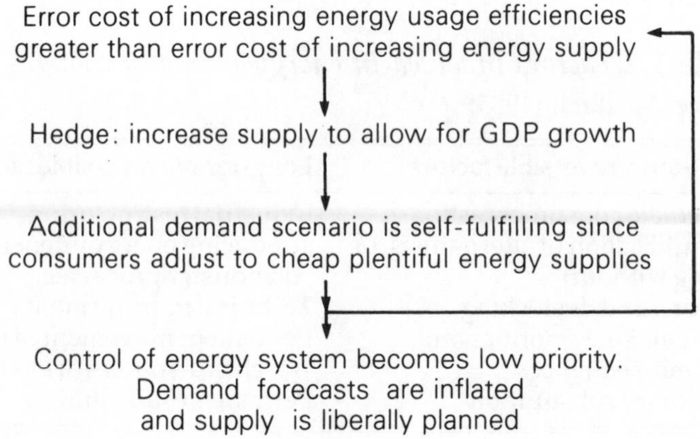

society beyond a state of equilibrium into an overbalance that presents itself as a fresh challenge'.

Global societies of all political persuasions are therefore unified in their quest for the ultimate technocracy. In fact, in many ways economic indices are taken to indicate the relative prosperity or poverty of particular political ideologies. The oft repeated call for greater energy usage is in reality the cry of societies without options (Higgins, 1978).

As Mardon *et al.* (1978), have shown, progress is identified with anything that increases the amount of energy and materials that people control. A basic paradigm in Western society is that energy and materials should increase each year; this is the design assumption. Now it must be realised that when the basic tenets of our civilisation are defined by experts, or a self-accrediting elite group, then inevitably there is a decline in freedom and a restriction of rights. This is the case at this point of time in Australia. Our industrialised society is now organised around what are termed professionally defined needs (Illich, 1973).

Our current society is therefore in the process of perpetuating paradigms involving three basic illusions. Each citizen now believes that he/she is a client requiring a saving service available only from the expert. The first illusion is that people are born to be consumers. The second deception is that economic models can ignore use values – exchange values can only replace use values in very limited contexts. The last illusion relates to the lack of distinction between personal and standardised values: extreme educational specialisation renders the citizenry generally impotent (Illich, 1973).

The belief that 'clean' abundant energy is a panacea for all social ills is a political fallacy. Energy and equity only 'grow' concurrently to a certain point – the threshold of technocracy – (Illich, 1973). The industrial world lurches from one energy crisis to the next; these are not crises at all but simply the manifestations of the contradictions implicit in the joint pursuit of equity and industrial growth. Equity and industrial growth are mutually exclusive: industrial growth pollutes and equity is consequently lost. The call for more energy is nothing other than a symbol of a society without options:

- burning petrol/oil products gives greater mobility in polluted air;

- burning coal for electrical power gives us ecological degradation;
- generating nuclear power leads to catastrophe;
- the mechanisation of agriculture means that we use forty joules to generate one (Higgins, 1978).

For any sustainable economy there is a need, firstly, for a renewable resource base, and, secondly, for a scale of *per-capita* consumption and population that is within the sustainable yield of this base. In terms of defining future policy there are really only two postulates (Schumacher, 1974):

(*a*) society can allow technology to devote itself to increasing the total throughput with no recognition of a limited biophysical budget or throughput – this is the economic growth scenario, or

(*b*) society can define a sustainable yield throughput or aggregate biophysical budget constraint and live within it, allowing technology to devote itself to increasing efficiency with which we use the given throughput – this is the steady-state or negative growth scenario.

Economic growth cannot last much longer as a paradigm within the Western culture. Restraints of energy supply, pollution problems and the ever-diminishing returns of technology will eventually prevail to counter the growth philosophy. Ecology, a discipline birthed in the sciences and transferred to politics after the Second World War (Rens, 1982), is now in the process of returning to its birthplace with a consequent awakening of awareness of the futility and destructiveness of eternal growth, development and progress.

The whole argument for the justification of growth rests on three basic precepts as defined by Evans & Atkins (1979, p.143):

- the concept that more goods and services *per capita* per year is equatable to a better life – an irresistible argument for third world citizens;
- economic conventions whereby growing economies are less difficult to manage than steady state situations;
- fear of losing out to one's global neighbours.

The problem this engenders has been well defined by Diesendorf (1979, p.5): 'To produce economic growth, a high standard of living and hence a high quality of life, energy consumption in the rich countries must double, or even triple, by the year 2000. What existing energy industries can be expanded and what new energy conversion technologies can be introduced to fill the gap?'

The balance needs to be defined between the physical environment (life and material support) which involves the take-up of energy and the absorption of waste products, and the ecosystem sustainability involving such concepts as the quality of life and human health. Demand satisfaction economics is overly concerned with the former and ignores the latter. It may well be that this is simply because the former is easily quantifiable and therefore amenable to 'scientific' solution; the latter is not. Most energy policies embody the assumption that the community is a resource capable of manipulation: energy policy is therefore divorced from energy considerations *per se* and it becomes a mere political activity (Higgins, 1978).

Unfortunately the concept of conservation is commonly equated with preservation; this is completely fallacious. Conservation ideally relates to the long-term management of natural resources, of which energy is part, for the benefit, short and long term, of mankind. The management of resources involves study of the interaction of resources and policy is definable only in terms of a 'way of life', a set of moral principles.

There is a lack of morality currently as evidenced by the profligate way that energy is used; food supplies are now effectively made of petroleum. Prior to industrialisation people survived without massive inputs of energy to the 'economy', so why should life be impossible without huge inputs of more energy? A major conflict here rises between the interests of nations and the interests of the large multi-national corporations, many of which are energy based. The maintenance of viable economies is increasingly being geared to the price of oil. This has great repercussions throughout the third world (and fourth and fifth worlds) which are now experiencing problems of firewood provision and other cheap cooking fuels.

2

The changing historical perspectives of energy supply

Our planet, the Earth, maintains a delicate thermal balance that makes life possible. This system of interdependence of living things, flora and fauna, is known somewhat affectionately as the 'balance of nature'. In essence it is a self-regulating and self-limiting system in that the biomass turnover can never exceed the conversion rate of incident solar energy. This simple relationship has defined the population levels of all living organisms with the exception of the human race.

The use of fire and the ability to make tools (commonly now eulogised in the term 'technology') have allowed mankind to break out of the population regulating system. The usage of ever-increasing amounts of energy, generally in the form of stored solar energy (gas, coal, oil), has given rise to the current status of industrial and technological man.

The scientific and industrial revolutions of the 19th century have set off a seemingly accelerating growth in the energy consumption of the global population. In the UK the use of the steam engine in the 18th and 19th centuries saw the first major increases in the *per-capita* use of energy. Despite major advances in the efficiency of energy use the 20th century technological man with all of the mechanical–scientific–technological innovations is using energy at a rate of some 22 times that of 'palaeotechnic man'(some of the poorest of the third world countries). Since man ceased to live in ecological balance with the environment the global population has increased some 1000 times whilst the energy consumption has risen some 100 times. It should be noted that the rate of solar-energy conversion has remained static. Apart from the fact that the energy consumption increases have occurred very selectively with respect to the nations of the world, hence the distinction of 'developed' and 'less developed', the majority of the energy expenditure has come, and is still coming, from the conversion of stored solar energy, *viz.* the fossil fuels.

The increasing utilisation of energy has steadily increased as a result of both population growth and increased energy usage *per capita*. Initially consumption patterns were influenced by the existence of local supplies, but in terms of overall consumption there have been major shifts in primary fuel usage as is shown in Figure 2.1. Since 1850 the world has passed from a predominantly wood based energy supply to coal in the industrial revolution to the current fuel oil base. During this period of time the world annual energy production has increased some 15-fold to the current level of about 310 EJ, Table 2.1. Most of this increase has been due to the industrial growth of the advanced capitalist countries such as the USA, the Western European nations and Japan. It has been the intent of governments of various political persuasions to raise living standards that has brought about the large increase in energy consumption.

From analytical perspectives energy for the use of man can be thought of in two ways. Firstly, there is an energy requirement in terms of food, water and air which might be considered the basic necessities of life. Secondly, there is the process of energy production to procure the material 'necessities' of industrial life.

Figure 2.1. World energy usage patterns. Source: Saunders (1976).

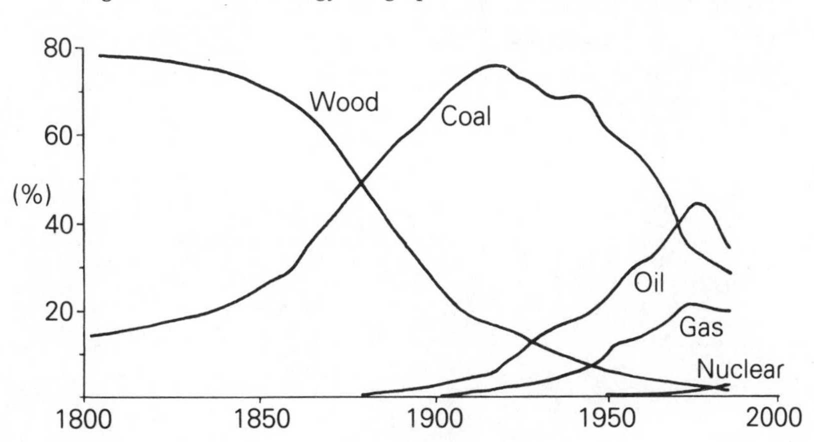

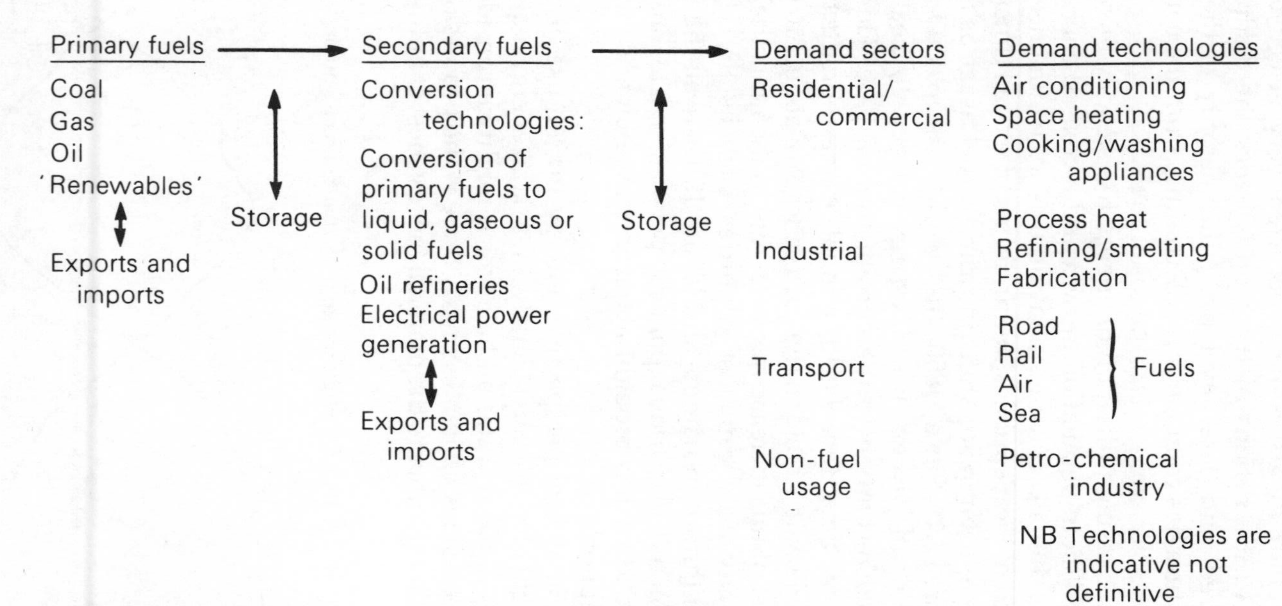

Figure 2.2. Block diagram for the Australian energy system. Adapted from Musgrove *et al* (1983).

The modern high energy using societies have an enormous capacity to control and transform the physical environment, locally, nationally and globally. An outline of an industrial society energy system is given in Figure 2.2.

The actual usage and impact of industrial technology has now penetrated to almost all human societies; however, the geographic distribution of the end use of that energy is greatly unequal. As the industrial and technological revolution has progressed, two factors of global importance have emerged. The first is the depletion of fossil fuel reserves and the second is the impact of waste discharges into the environment. Much of the consumed energy ends up as waste heat or waste gases and unwanted particulates (dust). The question or problem of energy supply is far more intractable than, say, that of a century ago: in a global context the question of equity has both international and national connotations.

Within the industrialised or developed nations the quest for energy supplies has been geared not only to transport but to modes of supply that can be adapted to centralised electricity networks. This has provided the incentive for the development of alternatives and innovation; there are advocates for solar 'power', nuclear fission and ultimately fusion, tidal energy, wind power and others.

The argument for nuclear fission commonly follows the theme that 'fission' is cheaper, far less polluting, and more efficient

Table 2.1 *Growth in global energy consumption*
Sources: *UN Statistical Year books, BP Statistical Review.*

Year	World annual production 10^{18}J	Cumulative world production 10^{18}J	Cumulative world population 10^6	Energy production per capita 10^9J
1900	22		1571	14.0
1925	45	980	1965	22.8
1950	77	2370	2501	30.5
1960	130	3400	2986	42.8
1965	162	4130	3288	48.5
1970	215	5080	3610	58.7
1975	251	6030	3890	63.3
1980	289	7501	4720	61.2
1985	310	8981	4838	64.1

than coal or wood. These arguments are only valid if one assumes a technology that works at peak efficiency continuously. Arguments against nuclear power include references to the potential hazards of radiation, waste disposal, storage and transport, and proliferation of nuclear weaponry. A major problem is that nuclear power production has the unprecedented effect of threatening all living matter. In terms of storage of wastes the major consideration has been the inability to guarantee long-term human political stability.

In the modern world the trend toward urban conglomerations means that energy questions, demand, supply, resources, have to be considered in an urban context. Until recently technologists have been the main determinants of energy use modes. The formulation of energy policies is a phenomenon of the past 20 years. Energy policy considerations cover aspects of the environment, urban planning, transport, housing, industry and human behaviour patterns. The main focus has naturally been on the high energy consuming 'advanced' technological societies that have emerged since the Second World War.

One of the major impacts of the increasing usage of energy is that it has allowed a large increase in the productivity of labour. This 'structural economic change' has involved the large-scale displacement of people, hence the surge toward urbanisation.

2.1 Utility of energy supply

Energy usage efficiencies can be calculated in various ways. Until fairly recently the technical efficiencies were calculated according to the first law of thermodynamics. A more useful measure of efficiency is defined by the second law of thermodynamics, where

$$\text{first law; efficiency} = \frac{\text{useful work done}}{\text{energy input}}$$

$$\text{second law; efficiency} = \frac{\text{useful work done}}{\text{maximum possible work done}}$$

As shown in Table 2.2 the 'second law' efficiencies can often be of the order of one-tenth or less than those of the first law. This means that they can show more readily where energy savings can be made. Utility of energy supply is in reality a problem of quality of energy available; entropy is the key. It is not only quantity of energy available that is a major consideration, it also relates to quality available (Lustig, 1979).

The flow of energy in an economy is shown in Figure 2.3. Since 1950 governments have believed in growth, commonly as measured by GNP or GDP. It is probable that there is no single reason why nations have been devoted to the idea of growth as a desirable end in itself. Possibly the three arguments given below are major considerations. The notion that more consumer goods and services *per capita* is equivalent to a better way of life is virtually an irresistable argument to someone barely managing to survive. Why this theme should permeate through advanced industrial/technological nations is the subject of much debate.

Secondly there is a concept that a steady-state economy is more difficult to manage than a growing economy. This theme seems to permeate most conventional economic thinking. Neither of these themes seems to impact greatly on the populace, essentially, we assume, because both are intangible and invisible. Lastly there is the 'fear of missing out'. The fear that one group, community or nation, if it does not grow, will be engulfed by those 'growing' around it (Evans & Atkins, 1979).

The basic types of available energy are shown in Figure 2.4. The current consumption relies very heavily on the fossil fuels, and in particular coal, oil and gas. The actual amount of primary energy consumed compared with total stocks is barely significant; however, compared with readily available low-

Table 2.2 *A comparison of first and second law energy efficiencies*
Source: Lustig (1979).

Energy use	First law	Second law
Room heating (gas)	0.6	0.03
Electrically heated water	0.75	0.015
Ditto, including power station losses	0.25	0.015
Gas heated water	0.5	0.03
Boiler	0.8	0.225
Air-conditioning	2.0	0.045

entropy stocks this consumption pattern is non-sustainable in the long term.

Of all energy resources, solar energy is perhaps the most tantalising and challenging, yet the most elusive and frustrating. The energy available from this source is immense, but the problems associated with its use relate to its low level of intensity and its temporal and geographical variability. The total solar radiation intercepted by the Earth's diametral disc is approximately 5.5×10^{24} J *per annum*: 70% of this radiant energy reaches the Earth's surface, the rest being reflected and dispersed. When reduced by the fraction of daylight hours, the total annual energy *potentially available* is nearer 10^{24} J. Even so the amount of solar energy reaching the Earth in just three days is equivalent to the *total* known world fossil fuel reserves.

The development and exploitation of solar energy has been restricted by its diffuse and intermittent nature. For commercial use on a continuous basis, collection and storage of the energy is necessary. While the energy costs nothing, the costs of collection and storage are high.

Figure 2.3. Energy and money flows at national levels. Adapted from Evans & Atkins (1979).

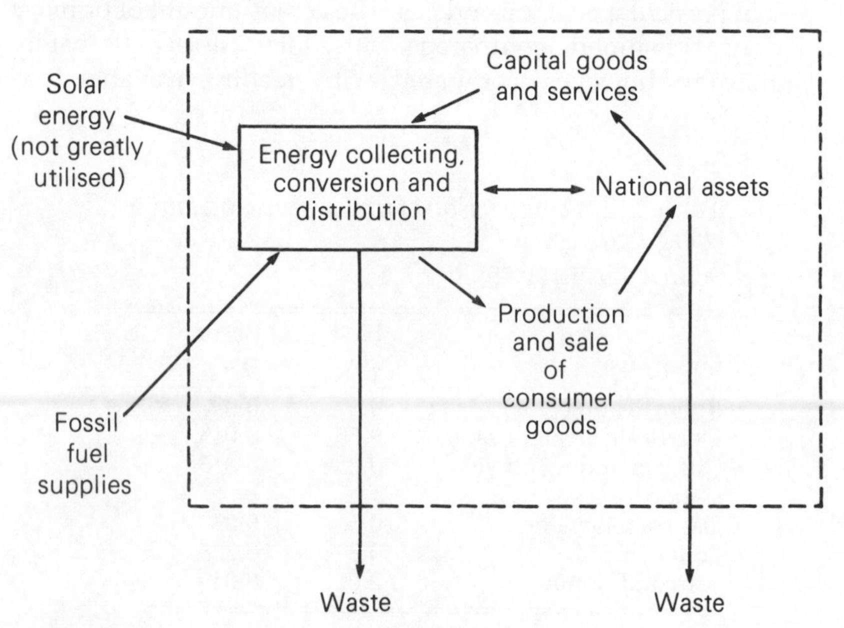

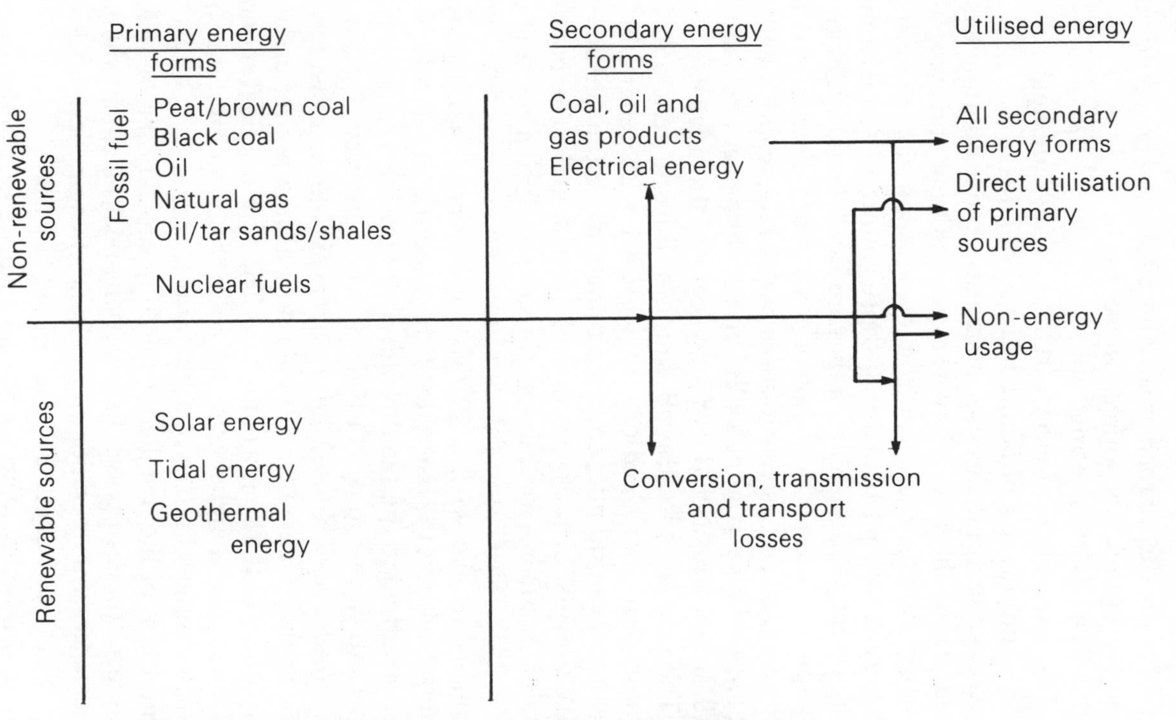

Figure 2.4. The basic types of available energy. Adapted from Grathwohl (1982).

There are three broad areas where solar energy could become a significant energy source. These are:

- low grade heating;
- generation of electricity and high grade heat;
- production of liquid and gaseous fuels.

In the attempt to develop alternatives most attention has been focused on the Sun. However, gravitational and natural heat energy sources could contribute, substantially, to easing the 'energy crisis', particularly in those countries that do not have long periods of sunshine, or that are locally endowed with these energy resources. Tidal power and hydro-electric energy are already used extensively in some parts of the world. Geothermal energy is also commercially viable; however, its large-scale utilisation is restricted by its geological infrequency.

The prospects for the utilisation of tides to generate energy, and in particular electricity, vary greatly with geography. Total theoretical world tidal energy dissipation is estimated at 3×10^6 MW. Of this, only 2% is currently considered to be potentially usable. Estimates for the energy recovery efficiencies range from 8 to 25%. Therefore, future output when fully developed is unlikely to exceed 2×10^4 MW *per annum*. Harnessing this energy requires the rising tide to be impounded behind a barrage and returned to the sea through water turbines as the tide falls. The tide varies on a 12.4 h daily cycle, superimposed on a 14 day lunar cycle. Hence, the energy available varies widely from day to day. Generally, the energy available is proportional to the square of the height of the tide. The simplest tidal generation schemes use a single basin and provide an intermittent output, which changes both in quantity and timing daily. Only a small proportion of the installed generating capacity can be relied on as a firm source of power to meet daily peak electricity demands. The successful operation of the Rance Scheme in France, since 1966, has confirmed the reliability of tidal power.

The Kimberley region in Western Australia contains nearly all of Australia's potentially exploitable tidal energy. Generally, the tidal range is 9–12 m throughout much of the region. There are a number of basins and inlets, with narrow entrances suitable for damming. These locations compare favourably with some of the Bay of Fundy locations. The Walcott Inlet in the Kimberleys has a potential maximum power output of 1254 MW and an annual energy output of 3.95 TWh, compared with 240 MW and 0.54 TWh for the Rance.

Wave power is a secondary form of wind power. It has been suggested that it may be possible to construct floating structures that will extract up to 90% of the displacement energy of wave motion. This would be converted to electrical energy at efficiencies of 60–70%.

A wave power station would be enormous, but technologically simple. The station would probably be between 0.5 and 1 km long, if it is to be stable in heavy seas, and it would also require anchoring to the sea bed. Alternatively, it could be towed out to sea, and allowed to move gradually along with the waves. The cost of a wave power station would be quite high; and figures in the range of $US 500–2000 per kW have been quoted.

The slow decay of naturally occurring radioactive elements (uranium, thorium, potassium) in the Earth's crust has generated large quantities of heat within the crust. Geothermal energy is currently tapped by utilising naturally heated water and steam from such hot regions in the Earth's crust. Plans are at hand to attempt to utilise the energy of hot dry rocks at depths of greater than 10 km. High-enthalpy geothermal fields have been developed in New Zealand, the USA, Japan, Mexico, USSR, Italy and Iceland. These fields are usually associated with regions of youthful volcanism, crustal rifting, and mountain building. One of the world's major geothermal and volcanic belts is the Circum Pacific margin.

Current world geothermal generating capacity is estimated to be 1300 MW, of which 800 MW is located in New Zealand, the USA, Mexico and Japan. Additional plants under construction in these countries, plus developments in Indonesia, Phillipines, Taiwan, Chile and El Salvador, could add an extra 800 MW to the Pacific region's capacity.

Geothermal systems so far developed fall into two main categories. A hot-water system is one in which the rocks are saturated with hot water, commonly at temperatures considerably above 200°C. Steam is flashed from the water by reduction of pressure and then separated from it. A vapour-dominated system is one in which the steam phase is continous, though hot water may be present in pore spaces. Wells drilled into a vapour-dominated system produce undersaturated steam.

Water geothermal systems are by far the most common. They range in magnitude from single warm springs to large systems like the Wairakei field in New Zealand which has a cross-

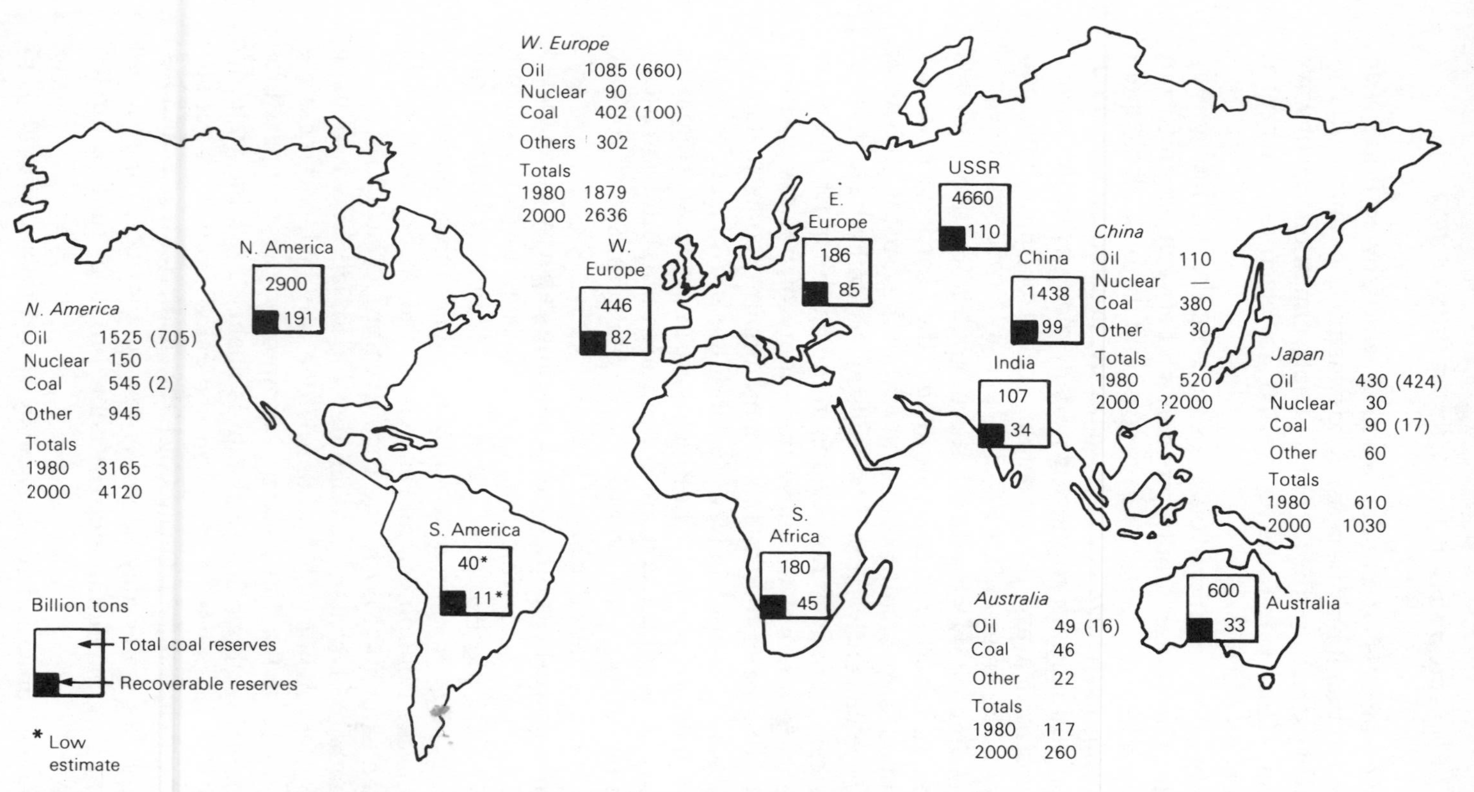

Figure 2.5. World coal reserves. Tables relate to total annual energy needs in million tons of coal energy equivalent. Bracketed figures are imports. Source: Adapted from Cunningham (1981), *New Scientist* (07,05,81, pp. 338).

sectional area of 35 km^2, or to large systems like that of the Salton Sea which has little or no surface expression. Most contain and discharge saline waters, though compositions are affected by the types of rock through which the fluids pass and by the superficial processes of oxidation and boiling.

Alternative fossil fuels are likely to be used in any one of three ways in the future. They can be used directly as a fuel; converted to more attractive liquid and gaseous fuels; or used in new technologies for electricity production. As the most pressing short to medium-term problem is a shortage of liquid fuels, particularly transport fuels, much attention is being directed at technology which will hasten the extraction of liquid–gaseous fuels from coal and natural gas. In addition, other non-conventional fossil fuel resources such as tar sands and oil shales are likely to be developed.

2.2 The politics of energy supply

The great bulk of the world's energy resources are situated in the Northern Hemisphere: more specifically about 90% of the proven reserves of hard coal are in N. America, USSR, China, Germany and the UK; about 40% of proved soft coal reserves are in Germany and the USSR. Over half of the known reserves of oil and gas for the Western world are in the Middle East, although there are indications of very large resources in the USSR and Mexico. The distribution of the world coal reserves is shown in Figure 2.5.

On a *per-capita* basis, N. America, the USSR and Australia appear well endowed with energy resources, although most of the resource for Australia is in the form of uranium ore, a resource of uncertain potential. The USA has a better balance, although still heavily reliant on nuclear 'fuel'. Japan is markedly deficient in *per-capita* terms, a feature it shares with the majority of mankind. Ironically, in the third world it is not a lack of nuclear, petroleum, gas or coal fuels which is cause for concern, but a lack of wood, which still forms a basic fuel need for much of humanity.

In terms of the geographic distribution of world energy reserves about 30% of the readily available low-entropy energy is in Russia and China; about 45% in N. America; and a very large part of the remaining 25% represents Australian uranium and coal.

Table 2.3 provides a review of current world energy reserves, with a particular emphasis on crude oil reserves. The data shown are a compendium of reserve estimates from various sources, and serve to show the degree of variation in published data. It should also be borne in mind that in any comparison of energy resource data, the terms 'reserves', 'resources', 'potential', 'proven' and 'possible', are generally ill-defined. Also, the discovery and assessment of coal is cheaper and more readily accomplished than for oil or gas; therefore it is probable that currently more is known about the extent of coal reserves than any other energy form.

The world's economically recoverable reserves of coal, both black and brown, amount to about 410 times current consumption rates. Likewise for oil, the static depletion rate is equivalent to 33 years, and for natural gas 44 years. If due

Table 2.3 *World energy reserves (10^{21}J)*

Energy source	Range	Proven reserve	Potential reserve
Oil	L	3.5	11.0
	H	4.8	16.5
	A	3.9	14.0
Coal	L	2.9	32
	H	65	285.0
	A	17.1	140
Uranium	L	218.5	265
	H	252.0	350
	A	235	309
Oil shale tar sand	L	4.4	12.5
	H	5.7	23
	A	5.0	18
Natural gas	L	1.3	3.6
	H	2.5	14.0
	A	1.6	8.0
Fusion			10^{10}
Solar			10^{3}
Hydro			6

L = lowest, H = highest, A = average estimates

allowance is made for potential reserves, then these ratios are considerably extended. By contrast, if the world should continue to increase its energy usage at a rate of about 5% *per annum*, these ratios are considerably reduced. These present reserve ratios can only be maintained by the discovery of new reserves to balance the yearly consumption. It should also be borne in mind that the relative energy usage ratios, in energy terms, are 1.8:1.5:1 for oil:coal:natural gas. The world economy is currently hooked onto the energy sponge we call oil, a dependence highlighted by the recent energy 'crises'.

The OPEC cartel holds about 68% of the reserves of world oil. The overall dominance of OPEC may not be as great once the full potential of the Mexican oil fields are realised. Recent estimates quote 250×10^9 barrels proven with a probability of double this reserves figure. This could substantially alter the balance of 'oil power'.

The major oil consuming nations are the highly industrialised economies of the world, which annually use about 80% of world production. By contrast, the 'non-industrialised countries' use a mere 16.0% of total world production. The political distribution of the world energy resources and end usage is given in Table 2.4. This well shows the dependence of the industrialised sector (IS) on imports of crude oil. In the short term, energy reserves in the IS, which includes Australia, are strictly limited, but for the long term, with the alternative inputs of coal and oil shale, these reserves are comparatively much better. However, for the short term of the next 10 to 15 years, the dependence of the IS upon oil as a source of energy will remain critical to the maintenance of the economies of those countries. Almost half the primary energy consumption in the IS is oil based.

In terms of coal the dominance of the OECD (Organisation for Economic Co-operation and Development) and the COMECON (Council for Mutual Economic Aid) political groupings are well known. Between them these two account for something like 77% of recoverable coal reserves and some 84% of resources.

In terms of oil, the distribution of petroleum over the economic–political groups is again uneven. The OPEC countries contain some 69% of the proved recoverable reserves, the Eastern bloc has about 14%, and the OECD about 9%. Natural gas reserves are somewhat more evenly distributed amongst the political groupings. The OECD has some 24% of reserves, OPEC about 22%, and the centrally planned economies some 34%. Overall the OECD is still in a very strong position to dominate and control the global energy supply. Provided the technological and industrial nations can get 'unhooked' from oil in the next ten years there is no reason to suppose (barring global war) that they will not continue to dominate consumption patterns globally.

Table 2.4 *Political distribution of world energy resources and end usage, 1985*

Data Source: *BP Statistical Review*

Fuel	IS	LDC	SOS
	% in each world sector		
Oil	12*	76	12
Natural gas	16	39	45
Coal	53	1	46
Uranium	85	5	10
Oil shale	70	28	2

Sector	Import	Export	Net export
	World oil trade by sector (10^{18}J)		
SOS	1.14	3.22	2.08
IS	30.84	4.47	−26.37
LDC	4.82	29.44	24.62

*7 (proved primary energy resources)
SOS = state owned sector (Communist Bloc)
IS = industrial sector (EEC, Japan, USA, etc.)
LDC = lesser developed countries (third world, Middle East, etc.)

3

Projected global and Australian energy requirements

As Bunyard (1976) has shown, projected energy demands on the basis of increasing industrialisation must inevitably lead to a state of large energy shortfalls in terms of supply. Limited capital, limited material resources and lack of skilled manpower all combine to produce this situation. In addition, the nuclear industry effectively is an actual consumer of energy rather than a net provider. Lastly, even if the energy industries do manage to come up with the extra energy, there is absolutely no guarantee that the consumer will be able to afford it. Bunyard (1976, p.100) writes:

> 'The sensible solution is to turn right away from the high energy society and look very hard and seriously for ways of living well on what we really can afford from our own fertile land. There can be no denying that to achieve a non-consumer society while maintaining a high standard of social care for everyone, including the old and sick, will necessitate something of a social revolution'.

Are the days of the industrial society really numbered?

Ironically in many industrialised countries it is the energy policies, both implicit and explicit, that define the major determinants of the level of industrialisation and social consumerism. Energy policy defines the level of economic activity, which in combination with population numbers defines the rate of economic growth – a major determining index. Energy policy therefore directly defines the potential for a variety of energy resources and their utilisation paths. This also includes environmental aspects and conservation stratagems, both are discernible only in the context of an energy scenario.

The commonest theme for demand forecasting is what Dick & Mardon (1979) call 'extrapolationitis'. In the majority of cases linear extrapolation from past, commonly recent past, consumption trends is utilised to forecast future demand. Morally this is reprehensible, statistically it is insanity, and regarding policy it is completely unjustifiable. Dick & Mardon are less volatile, they simply describe it as 'hazardous'.

The two authors (Dick & Mardon) describe a better methodology utilising the concept of logistic curves to define potential demand. They show quite clearly that the energy consumption in Australia is reaching a condition of saturation. Perhaps the declining growth projections of the forecasters for the Australian scene reflect a gradual recognition of this point.

Exponential or linear projections can only be justified on the basis of growth, itself supposedly a characteristic of a society which is progressing or developing. Errors of this sort can be exceptionally costly, to whit the excess generating capacity of the Central Electricity Generating Board in the UK. This

Figure 3.1. Total primary energy projections – Australia. (*a*) Institution of Engineers (1977*a*); (*b*) Institution of Engineers (1977*b*); (*c*) DNDE (1978); (*d*) DNDE (1981); (*e*) DNDE (1983); (*f*) Dept.Resources and Energy (1 Energy intensive case, 2 Base case, 3 Conservation case). Updated from Stewart (1983).

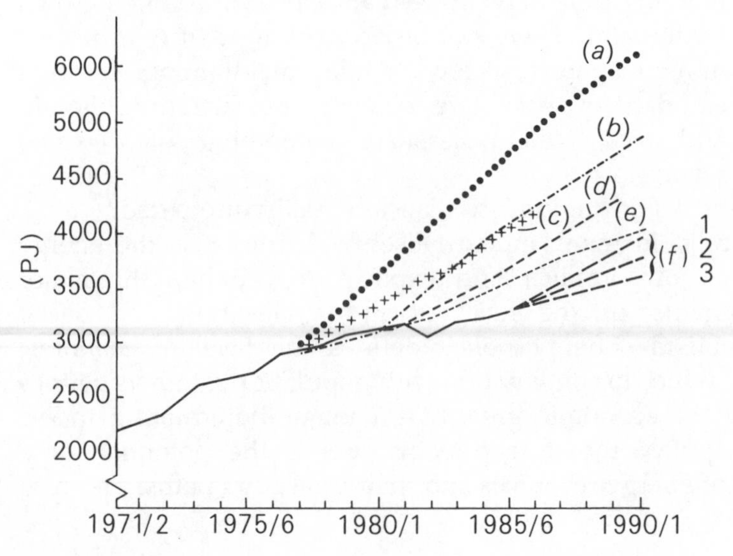

generating capacity was developed on the basis of linear extrapolation, Dick & Mardon (1979).

Examples of extrapolationitis are shown in Figure 3.1. In all cases shown the projected demands are nothing other than linear or quasi-linear extrapolations of recent historical growth. Thereby the actual dynamic of the energy situation is totally by-passed and the ethic of growth perpetuated.

The definition of a future energy demand must reflect the inter-relationship and interplay of various parameters, namely

(*a*) availability of energy forms;
(*b*) potential supply shortfall;
(*c*) problems relating to the fuel mix;
(*d*) perceived needs for substitution;
(*e*) conversion costs (dollar and energy);
(*f*) export/import balances;
(*g*) non-technical barriers to proposed energy policies.

Energy policies themselves will therefore reflect

(*a*) the 'level' of economic activity;
(*b*) the population, current numbers and the rate of increase;
(*c*) the technology (in-place and required) for each energy option;
(*d*) conservation and environmental problems.

It is doubtful if any of these parameters are taken into account when forecasting is done on the basis of linear extrapolation. With reference to Figure 3.1, even the most conservative prediction allows for a 30% increase on 1980/1 levels of primary energy consumption for 1990/1: the most optimistic allows for a doubling of energy consumption. Both of these proposed scenarios represent a huge demand on capital, resources and the environment, an imposition that is not necessary and not attainable. Until recently the rate of primary energy demand growth in Australia has been about 5.5% *per annum*, a doubling rate of some 13 years. This is thermodynamically a most inefficient economy.

By contrast the aim of a sustainable energy society, apart from being a thermodynamically efficient economy, is one in which, firstly, energy demand growth rates would be much lower than now, and, secondly, that most of the energy supply would be from renewable sources. Therefore,

$$E_p = E_u + E_w$$

total primary energy input = total useful energy + total wasted energy

represents an energy economy for which E_p is optimised to give an E_w low and an E_u high. Conservation measures tend to optimise the ratio of E_u to E_p (Roby, 1979).

Energy forecasting and the resultant policies represent a choice from amongst several alternate scenarios: they are not an end in themselves although they are often taken to be. The objectives of energy policy should be to provide sufficient energy to sustain in perpetuity a coherent society of high quality with substantial diversity and flexibility of life options for its individual members. Given this dictum should the approach to energy policy therefore be

(*a*) one of *laissez-faire*, whereby the market is taken to be the best indicator of prices and options; or
(*b*) one of intervention whereby it is assumed that government can do better than the market?

To date Australian energy polices have been a mix whereby government attempts to 'help' the 'market' so as to optimise profitability. These options must be decided against the backdrop of the macro-thermodynamic considerations, whereby

(*a*) more and more fossil energy is lost to man; and
(*b*) the overall efficiency of the energy system is decreasing.

3.1 The science and technology of energy

Following the 'oil shock' of 1973 many governments of industrialised nations went on a search for alternative energy sources. Many of these alternatives relied heavily upon the 'science and technology of energy': in other words, the successful implementation of these energy sources was reliant upon the application of science and technology. A few of them are discussed here. The most imaginative perhaps embraces the harnessing of solar power.

Solar energy has been successfully used for low grade heating with solar water heaters, solar heaters and drying systems having been developed to the commercial stage, and is now in widespread use. About one-third of total energy requirements in industrialised nations are used to produce low grade heat (below 120°C), and approximately 50% of total electricity demand is used in this way. Virtually all of this energy could be obtained commercially from solar energy, using proven technology.

The immediate advantages of using solar energy in this way are:

- saving on electricity use (and the fuels used to generate it) without reducing the standard of living;
- creation of new industries, with employment in all major population centres;
- environmental superiority of solar energy.

In addition to commercially available solar heaters, substantial research and development is being directed to solar cooling. Solar cooling would find its main application in areas where there is an internal energy surplus which must be dealt with to secure a satisfactory environment. Solar cooling units have been commercially developed in the USA.

While low grade thermal applications of solar energy have reached an advanced stage of development, the long-term possibilities in other directions are extremely interesting. The energy the Earth receives from the Sun is very high grade energy from a natural nuclear reactor with an equivalent surface temperature of some 6000°C. This radiation contains a relatively high proportion of high energetic photons which can be utilised to generate electricity.

There are three main possibilities for generating electric power from solar sources. These are:

- *Thermal conversion*, where the energy is absorbed thermally and converted to electricity by conventional heat engine methods.
- *Photovoltaic conversion*, where the electricity is produced directly by the absorption of light in chemical cells.
- *Thermo-electric generators*.

About 2% of the Sun's incident energy is converted to wind energy. The energy available from these winds is proportional to the cube of the wind velocity. The total wind power in the Earth's atmosphere is estimated to be 300×10^{12} MW, of which about 25% is available over land. The usable fraction of this energy has been variously estimated to between 1 and 20×10^{6} MW. However, problems exist in respect of variable energy supply, and also there is the storm hazard: systems have to be designed to withstand very high winds. Wind energy is most abundant in winter (in contrast to direct solar energy), and its major attractions include zero fuel costs and, essentially, no adverse environmental effects.

Capital cost depends upon many factors, including
- annual mean wind speed of site;
- quantity production from a single source;
- power rating of unit;
- rotor diameter.

Wind resources are intermittent, time varying and generally unpredictable. The application of the derived energy therefore determines whether storage is required. The historical use of windmills required no storage. The small-scale use of wind power for application to horticulture, agriculture, grain drying, fish farming may be possible in most cases without the need for a storage system.

It has generally been considered in the past that storage would be necessary for large-scale application of wind power for electricity production. The techniques suggested include
- pumped hydro schemes;
- pumped air storage;
- flywheels;
- batteries;
- hydrogen production by electrolysis.

There are few technological problems in building windmills with up to 80 m diameter rotors and power outputs up to 1.5 MW.

Within certain areas of the world's oceans ($\leqslant$ 1600 km either side of the equator), solar radiation produces thermal gradients of up to 20 °C between the surface and deep waters. The Gulf Stream exemplifies the best vertical temperature gradient, with a variation of 21°C over a 1000 m depth.

Ocean thermal power is the ocean's greatest renewable energy resource, being daily replenished by solar radiation. This power source is capable of providing, on a continuous basis, 200 times the Earth's total power needs in the year 2000. Since the surface water temperatures of tropical waters never fall below 25 °C at any time it would be possible to operate sea thermal power plants at near capacity for 24 hours a day. The ocean covers 90% of the Earth's surface in this equatorial region. Some of the most promising areas for sea thermal development are the South Atlantic equatorial current area, the Gulf of Panama, Micronesia, and the north-west coast of Australia south of Jarva. Other areas with usable temperature variations are the sea around Hawaii, most of the Caribbean, the Gulf Stream off the east coast of Florida, the Gulf of Mexico, the Arabian Sea, the Indian Ocean, the East Indies, and the Atlantic Ocean off the coast of West Africa.

One of water's most important properties in this application is that it forms relatively stable isothermal layers – layers determined by temperature and density – which can be tapped for sea thermal power without significant disturbance. The combination of this property with the global pattern of ocean circulation forms the physical and environmental foundation for sea thermal power. The heat engines in sea thermal plants transform heat energy into the mechanical work of spinning a turbine. Generators are connected to the turbines to produce electricity.

Next century the world will face an increasing shortage of liquid fuels, particularly for road transport, aviation and lubrication. A number of processes that will produce liquid and gaseous fuels from solar sources are currently being investigated. One of the main processes being investigated involves the conversion of energy stored in plant materials (biomass), such as trees, crops, sugar cane, water plants and cassava, to produce fuels such as ethanol, methanol and methane. Other processes involve the conversion of algae and wastes (crop, urban, sewerage, animal, etc.). Hydrogen can also be produced by the photosynthesis of water or the chemical breakdown of ammonia. Biomass is currently considered a potential medium to long-term option, and biomass research is active in many countries.

Of the total solar radiation, only 3.1×10^{21} J is trapped by plants and converted into chemicals; 99.9% of all incident radiation is unused. Of the energy stored by plants each year, approximately 0.5% finishes up on our plates as food. Some plants have been used for centuries to produce useful materials, but energy yields have been relatively low. The overall efficiency of normal agricultural processes in temperate climates is *less than 1%*. Higher yields, of the order of 5–10%, have been achieved in idealised laboratory conditions, and with special crops.

The normal conversion efficiencies of high yielding tropical crops, such as sugar cane, cassava and elephant grass, are in the range of 0.8 to 1.6% over a 12 month period. On the other hand, the productivity of indigenous eucalypt forests is low, with average solar-energy conversion efficiencies less than 0.03%. These low energy efficiencies are a problem that must be overcome if biomass production is to be viable. Further, biomass fuel production will require relatively extensive tracts of good agricultural land, if we are to produce a substantial proportion of our liquid fuels from this source.

One process that has been considered for large-scale use, because its energy efficiency appears satisfactory and could operate on a continuous basis, involves the conversion of the cellulose, obtained from forests, into either ethanol or methanol. Both of these fuels can be blended, up to 20%, with petrol without disadvantaging the engine performance of existing vehicle stocks.

Ethanol is produced from the cellulosic portion of the wood by hydrolysis, followed by fermentation of sugars. There are several possible hydrolysis processes, but only a few have been commercially demonstrated. One of these is the Rheinau process, which operated in Germany during the 1950s.

The production of ethanol or methanol from crops such as sugar cane, elephant grass and cassava would be much cheaper than alcohol produced from forests. Sugar crops, such as beet, cane and sweet sorghum, are all possible fuel crops. Such energy crops are expected to produce fuels at approximately half the cost of alcohols produced by wood distillation.

For high yielding crops such as sugar cane, it has been estimated that an area of 3.1×10^6 ha would be required for the production of one-third of Australia's primary energy consumption. This area is approximately ten times the area presently under sugar cane cultivation. The present production of sucrose in Australia is about 2.8×10^6 tonnes *per annum*. This would convert to approximately 2×10^9 litres of ethanol with a fuel value of 43×10^{15} J, or 3% of current Australian demand for petroleum fuel.

The average Australian sugar cane yield is 25 tonnes (dry weight) per hectare per year. Production inputs are in the range of 7–17% of the fuel value of sugar, bagasse and leaves. Additional energy requriements for irrigation amount to 7.5% of the fuel value of the crop. Thus, for biomass conversion to liquid fuels, with an efficiency of 20% or less, there is not likely to be any net energy gain.

A number of technologies have emerged for the conversion of wastes to clean fuels. The use of solid and liquid wastes is an ideal and economic way of meeting primary, or supplementing, energy needs and at the same time disposing of the 'wastes'. The following process routes are being developed for energy recovery from wastes:

(*a*) direct combustion of solid wastes;
(*b*) hydrogenation of solid wastes to liquid fuels;
(*c*) pyrolysis of solid wastes to liquid/gaseous fuels;
(*d*) bioconversion of solid wastes.

There is an inherent attraction, both economic and environmental, in using these materials for the production of fuel. In Australia agricultural and urban wastes amounting to about 0.5×10^{18} J of stored energy are potentially a source of organic materials for synthetic fuel production. The main problems in economically utilising these wastes are collection and transportation.

Urban waste water also offers considerable potential as a source of energy. Urban waste water generally contains small amounts of solids (less than 1% by weight). These solids consist of inorganic chemical compounds from the water supply and many complex organic materials derived from human and industrially derived wastes. They are present both in solution and in suspension. The principal organic compounds are urea, proteins, amines, fats, soaps and carbohydrates, including cellulose. There are techniques of harvesting energy directly from urban waste water and there are indirect ways by which its necessary disposal can bring about savings in alternative sources of energy.

There are three principal unit processes for waste water works; these are:

- *Primary sedimentation* to separate the majority of settleable solids from the flow to produce the first major by-product of the system – primary sludge.
- *Biological treatment* to convert dissolved solids into settleable biomass by the development of a colony of microbiological organisms.
- *Secondary sedimentation* to separate organic solids from the process flow. This unit process is the source of the second major by-product of the system and this is called secondary sludge.

The heat value of raw sludge is about 1300 kJ/person/day. If the sludge is digested then the heat value is reduced to about 500 kJ/person/day. This energy can be utilised by the incineration of mechanically dewatered sludge providing an effective means of conversion of the sludge to an inert ash of relatively small volume which can be readily disposed of without risk to the environment.

The suitability of hydrogen as a substitute fuel has long been argued; its potential can be assessed from the following:

- On a volume basis hydrogen has approximately one-third of the heat value of natural gas.
- Hydrogen burns to produce water. Thus, formation of normal fossil fuel pollutants, such as carbon monoxide, unburned hydrocarbons, sulphur compounds and particulates, is avoided. Nitrous oxide formation can be reduced by low temperature combustion.
- With proper burners and settings, hydrogen can be burned in household appliances.
- Unlike natural gas, hydrogen undergoes 'flameless' combustion when passed (mixed with air) through a process plate filled with a catalyst.

It is doubtful that hydrogen will be seriously considered as a major energy source for considerable time to come. The reasons being the energy cost of production and the safety aspects of storage and distribution.

3.2 Energy paths

Energy policies for many industrialised nations embrace the twin goals of minimised oil imports whilst sustaining growth in energy consumption. The usual solution is reliance upon coal, oil and gas, and nuclear power. This is the so-called hard energy path.

The world's coal reserves represent more than 15 times the energy content of known world oil reserves, and coal is the one fossil fuel that is likely to remain in abundant supply, at relatively low costs, for the remainder of this century, and well into the next. It is one of the main alternative fuels available to bridge the gap from the current era of abundant oil to a future era of renewable resources. Coal will be even more important if nations are reluctant to implement nuclear power options.

Reserves of oil shale are concentrated in N. America (60% of total reserves), Africa (15%), Asia (14%), and Europe (11%). Reserves of lower grade oil shales (40–100 l/t) are estimated to be three to three and a half times larger than the high grade deposits (> 100 l/t) with 60% of these reserves in N. America and 30% in S. America. The largest oil shale reserves are in the 44 000 km^2 Green River Formation in central-west USA with reserves of 600×10^9 barrels of oil in place, occurring in beds at least 3 m thick, and with a yield of 113 l of oil per tonne of shale.

Tar sands (oil sands) are agglomerates of water and wet sand particles surrounded by a high viscosity bituminous skin. The bitumen content can vary from 1 to 18% by volume. Known major tar sand reserves are restricted to the Americas. To date, Canada, the USA, Columbia and Venezuela have outlined substantial reserves. The two largest deposits are in Alberta, Canada (710×10^9 barrels) and the Officina-Temblador belt in Venezuela (200×10^9 barrels). The next largest is the notably smaller deposit of Bemolanga in the Malagasy Republic (1.75×10^9 barrels).

Many of the petroleum products obtained from the refining of crude oil can be replaced by other energy forms. However, motor vehicle transportation is almost entirely dependent on motor spirit, and to a lesser extent distillate produced from crude oil. However, liquefied petroleum gas (LPG) is an economical alternative to motor spirit for a significant proportion of motor vehicles. It is also a valuable resource for the petrochemical industry, as a domestic energy source, and as a feedstock for petrol production.

Large increases in coal utilisation, however, would have a profound impact on society's attitude towards coal. It would require developing new infrastructure and new technology for coal mining, processing and use. Coal could ultimately replace existing energy systems, which are designed for oil and gas; however, it would force the world to confront the serious environmental issues related to extensive coal mining and burning.

Natural gas is a clean, convenient and versatile fuel, used for heating, cooking, power generation, transport fuel feedstocks and petrochemical production. Whilst world reserves are substantial, and are unlikely to be limited before the end of the century, the future role of gas an an energy source will be determined by the problems of transportation and distribution. The transport problem has limited intercontinental gas trade. The large-scale use of natural gas has been in markets that could be economically connected, by pipeline, to gas reserves. The expense of constructing costly pipeline networks can only be justified if gas reserves are large and demand is assured.

An alternative to pipeline transportation is tanker transport of liquefied natural gas (LNG). The technology for liquefaction, transportation and regasification has only become commercially available in the last decade. Gas is liquefied, carried in special refrigerated tankers at temperatures of −161°C, and regasified at receiving terminals. However, there are a number of problems associated with LNG. The process of liquefying the gas is energy intensive, and approximately 20% of primary energy is lost. In addition, another 10% can be lost due to 'boil off' during long distance transportation of LNG.

A soft energy technology is, according to Lovins (1977), one that is flexible, resilient, sustainable and benign. The distinction between the 'hard' and the 'soft' energy pathway is not typified so much by the quantity of energy used but by the structure of the energy system. The differences are therefore technical, social and political.

The first difference is that the soft energy path relies on the renewable energy flows within the Earth's biosphere, *viz*. Sun, wind, and vegetable matter. The reliance is upon what might be termed energy income rather than energy 'savings' or capital, i.e. fossil fuels, which are depletable.

A second difference is that soft energy paths are characterised by diversity. Hard paths are essentially mono-fuelled relying upon a massive input of one fuel type, commonly either oil or coal or both. Soft paths are fuelled from many sources each appropriate to local needs rather than a centralised power generation format. Decentralisation is therefore also a key parameter.

Soft pathways do not rely upon a highly sophisticated technology, the key being the application of appropriate technology as envisaged by Schumacher (1974). This is generally construed to mean that the required technology is unsophisticated; this is not necessarily true, the emphasis is upon accessibility and ease of usage rather than esotericism. Using an appropriate technology means that the fuel used and energy supplied are scaled to the end use purpose. There is not the reliance upon a centralised power scheme supplying energy for all purposes to all people in that region. Centralised electricity generating schemes are characteristic of the latter, small-scale generating schemes – solar powered, wind generated electricity units – are typical of the former. To follow this type of soft path therefore requires that the energy quality be truly matched to end use needs. This requires society to analyse fully what type of energy it needs, and where, in what form and when.

The main axioms of Australian energy policy, apart from the aspects of topic fragmentation and lack of choice regarding end objectives, relate to a policy of provision for projected demand. For example, the role of the individual motor car, and its dominance in society, is to be maintained and therefore fuels must be provided to ensure maximum mobility potential. Energy consumption is assumed to increase by a factor of double or treble current rates by the year 2000; within this projected trend electricity is expected to satisfy a much greater percentage. To this end the policies formulated by recent Federal and State governments tends toward such concepts as coal liquefaction (although recent oil discoveries and the current global situation have cooled this of late), the maximum possible development of mineral exports and the prospect of a nuclear industry by the year 2000.

The rationale for this is claimed by its proponents to reside in the wish to preserve the present life-style. However, ever-increasing consumption of energy does not preserve the current (prevailing) life-style, it alters it. For example, high energy scenarios tend to exacerbate the problem of individual transport, i.e. they encourage private vehicle ownership with all of its consequent social and environmental problems. The ideal of provision of ever-increasing energy demands ignores the real world situation.

Low energy futures must be the objective of energy policies since they follow logically from consideration of various aspects of the energy spectrum. Firstly, the world's oil and natural gas reserves are essentially finite and will be substantially depleted by the turn of the century. Secondly, there is at present virtually no substitute for oil and gas, being the only energy sources that are easily transportable and storable. Thirdly, other technological solutions such as breeder reactors, coal liquefaction, utilising enormous science/financial inputs will not be brought 'on-line' with sufficient speed to allow for continued growth in energy demand. And lastly, and maybe most importantly, the hectic dash toward an immediate high energy short-term future will doubtless foreclose on many options allowing for a smooth transition to a low energy scenario. The time scale can be significantly shortened for the transition to a conservation and renewable energy source strategy if high technology solutions are rushed into place.

In terms of energy policy, itself a political activity, there are basically two views. One relates to the upgrading of present and the developing of new energy sources. The concept is one of more energy availability *per capita*, this being thought to lead to the better social, physical and moral well-being of man. The other view is one that concerns itself with the concept of steady state. Requisite to this is a slow-down in the frenetic industrial activity with a concomitant shift down in socio-economic expectations and a change in life-style.

The advantages of soft energy futures are that they can sustain expanding populations whilst at the same time minimising environmental damage. Soft energy futures are also compatible with 'loosely' defined sets of goals and policy objectives.

One of the problems of implementing a soft energy future lies in the required transfer of control of energy supply from a central forum to the general population. Soft energy futures are the antithesis of hard energy centrally developed systems. Disaggregated energy supply does not necessarily mean a loss of provision or a loss of quality or standard of living.

The development of large centralised energy systems is relatively easy and does provide a fairly constant power base, but is vulnerable to breakdown and terrorism. A breakdown in any part of a disaggregated system does not disadvantage the whole system as it does for a central power source. However, with the shift toward the 'information society' there will be a lessening in energy intensity in the developed world.

4

Energy sources

It is interesting to note that today we are utilising a relatively scarce resource to fulfil our energy requirements, oil, and not utilising other more plentiful energy resources. Coal, which is far more plentiful than oil, is currently undergoing a revival as analysts realise that oil is a limited-term energy option. In terms of recorded human history the 'fossil fuel' age is representative of only the past few hundred years (including wood as an industrial fuel source). It is anticipated that the whole 'fossil fuel' age will last only about 1000 years, that is until depletion of fossil hydrocarbons occurs. The time scale perspective is given in Figure 4.1. For some energy analysts, especially proponents of a nuclear future, this represents the spectrum of the energy problem. Since fossil fuel sources are seen to be finite, a

Figure 4.1. World oil and coal production – the historical perspective. Adapted from Titterton (1982).

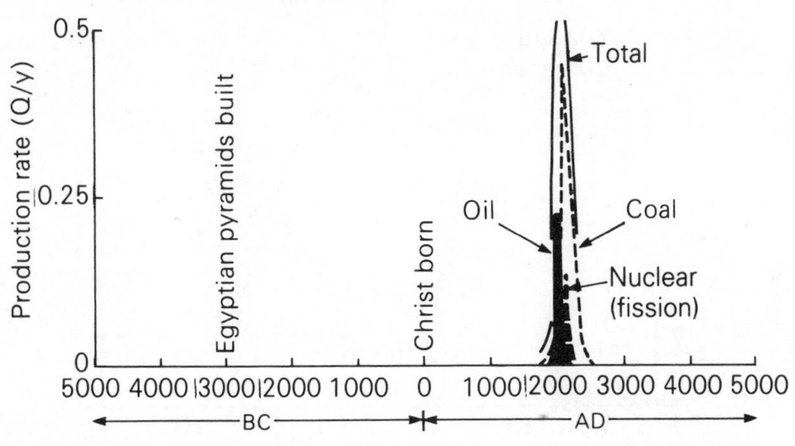

centralised nuclear power generated energy format is proposed as the only viable future for mankind.

From the turn of the century till the early seventies the consumption of oil had been increasing at the rate of about 7% *per annum*. The increase was fairly steady until 1973 when the 'oil crisis' quadrupled prices, see Figure 4.2. After the OPEC-imposed price rises the main fuel consumers, the Western industrialised nations, once again took up the growth trend. The 7% growth rate in fuel consumption was maintained until virtually the end of the 1970s. The combination of a world recession, the eruption of a Middle East war (Iran and Iraq), and an end to the era of oil-fired power station building, plus a trend toward conservation brought on by higher liquid fuel prices, considerably reduced the growth trend.

As Figure 4.2 shows the oil resource cannot supply a burgeoning world energy demand beyond perhaps the next 50 years. It is anticipated that oil production will peak around 2000 AD and thence go into decline. This means that temporarily at least oil supply will exceed demand; hence the current 'oil glut'. This should lead to some stabilisation of prices and perhaps even a falling in price over the short term. This phase will not last long, and by the turn of the century prices should rise once again as the resource supply begins to dwindle.

It is unfortunate in many ways but the two 'oil shocks' of the 1970s have, to all intents and purposes, destroyed the economic

Figure 4.2. World oil production and consumption. Adapted from Titterton (1982).

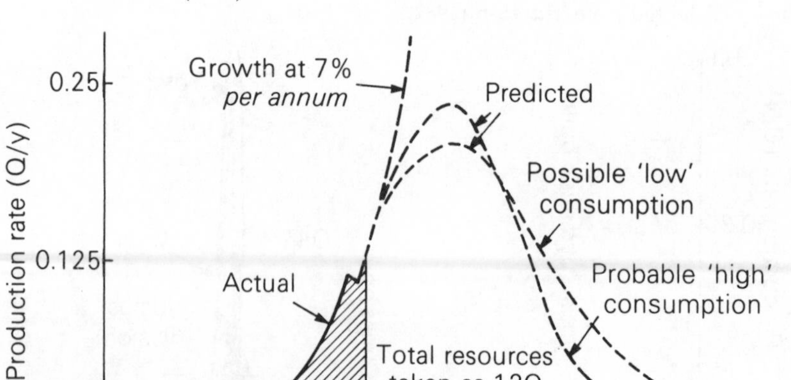

well-being of the industrial democracies. There should never have been an oil shock in 1973: there was no economic imperative, no lack of reserves. 'There was no pattern of necessity at all but, rather, a jumble of politicians' blunders, negligences, and timidities the connecting thread of which was the personal opportunism of the moment and the postponement of the inconvenient consequences to the next fellow's term.' (Anderson, 1984, p.xiv).

The current oil glut is the result of greed and necessity. OPEC cannot make or enforce oil policy or decisions because its own members cheat on both production quotas and prices. The greed comes because of the desire for a greater market share; the necessity because of the developing oil-exporting nations. The economics of the 1980s have already been determined by the oil politics of the 1970s.

A similar growth-decline pattern is expected for the coal resource, Figure 4.3. The actual time span of the production curve is much greater than for oil, and peak production is not anticipated until about the end of the 21st century. As the figure shows the growth in coal consumption has not been as smooth as that for oil. Since the end of the Second World War, however, an annual growth rate of some 3.5% has been generally achieved.

The curve in Figure 4.3 shows that for the next few decades at least coal supply is capable of exceeding demand, or at least

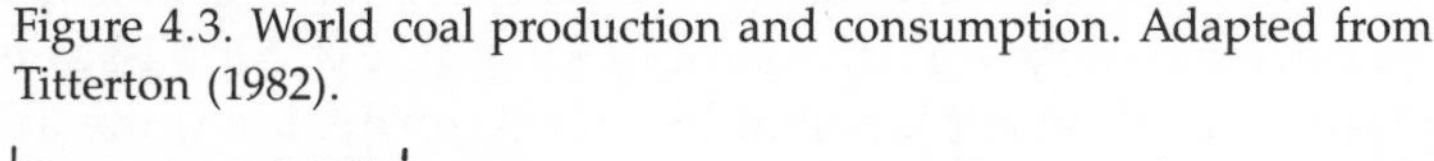

Figure 4.3. World coal production and consumption. Adapted from Titterton (1982).

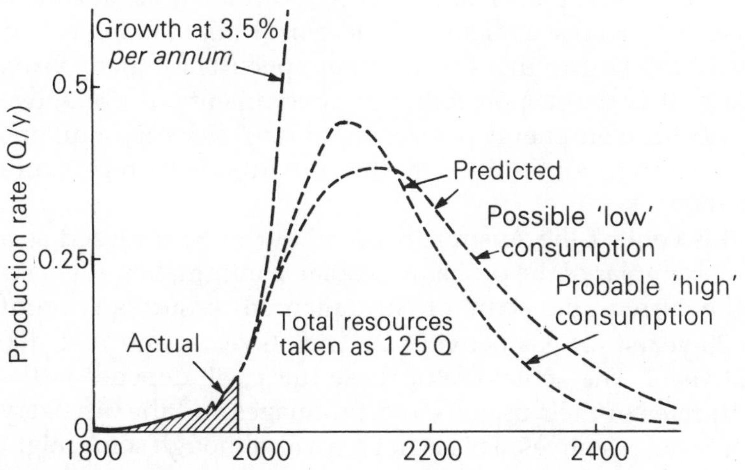

current demand. If liquefaction becomes commonplace then demand will increase significantly and the coal production curve will tend toward the picture as exemplified for oil. The actual coal consumption pattern will reflect the perceived consequences of the Greenhouse Effect.

These graphs show that world energy consumption is geared currently to the usage of non-replenishable fuels: coal, oil and gas. Now, although large stocks of these resources are available, the depletion rate is such that, given the supply–demand vagaries due to political considerations, an alternative fuel supply is a requirement for global societies of the 21st century and beyond. The argument therefore centres around whether nuclear fuel, which, dependent upon whether the source is fission or fusion, can be considered as either non-replenishable or replenishable, or 'soft' energy forms (solar, hydro, etc.), which are truly replenishable sources, should be utilised. A compounding factor in the argument is that nuclear power, disregarding the waste disposal problem, has a demonstrated capacity for large-scale power generation, whereas the 'soft' energy forms do not. For example, large power generating nuclear stations already exist (i.e. $>$ 500 MWe), whereas no such solar, wind or wave power station has even been built or shown to be technically or economically viable. This argument is countered by the soft energy path proponents who argue for an 'appropriate energy supply' format. What is meant by this is that large-scale centralised power generating plants are not really necessary for the supply of energy. Small-scale 'local' plants based on replenishable energy forms are quite adequate to ensure societal well-being of the global population. Vested interests will ensure that the argument persists for many years to come. It is the responsibility of governments to rise above this and formulate energy policies suitable to the local conditions and ensuring continuity of the environment for future generations.

In this context the Australian situation can be regarded as a typical exemplar of the problem. Available information suggests that the current 'life span' of Australian oil resources is about 10 to 20 years, for gas between 25 and 50 years, and coal, 100 to 400 years. The actual life of these fuels will depend on the growth rates of their usage, export tonnages and the discovery rate of 'new' resources. This is not a crisis, although some might term it such, since there is no particular turning point or decision moment. It is a continuum with changes in usage patterns occurring but very slowly. Oil, gas and coal stocks can be made to last longer than the time spans indicated above by more judicious usage as will doubtless occur when stocks are truly limited. However, for the long-term provision of a viable replenishable energy source, it is imperative that the declining years of relatively abundant oil supply be utilised to define and develop the technology for utilisation of such a fuel.

In the past few years there has been a relatively rapid development of solar technology in the developed nations as the impact of increased oil prices and realisation of the dangers of oil dependency have gradually been assimilated. Major advances in materials technology and in production techniques have helped to lower capital costs associated with solar collection equipment. This trend could well continue if and/or when solar markets become available in the future. But solar energy could have an important role in the global energy supply.

In the past few decades the main function of solar energy is likely to be the replacement of oil and gas as energy sources for heating up to 95°C. Commercially available solar equipment can be used for this purpose. It should be possible in the near future to extend this range up to about 150°C. The potential for solar-energy use is great, but even a modest goal of providing 5% of Australia's total energy demand by solar radiation within the next 25 years would require enormous capital investment.

There is less likelihood of solar energy being used to a significant extent for electric power generation. There are several reasons for this: the technology is not yet fully developed and the cost of equipment is not likely to be competitive with coal-fired power stations. There is, however, a small but important possibility for generating electric power by solar radiation in remote areas.

The production of liquid fuels from plants is likely to be looked at closely in competition with other possibilities of providing liquid fuels. The production of hydrogen and its use as fuel should not have a strong effect on the overall energy situation. However, should the cost and performance targets for photovoltaic cells be reached then the above forecast could change – the intermittent operation of the cells would not be a disadvantage here.

By contrast, the future for the nuclear industry looks a little less assured. Rising construction costs, falling oil prices, the decrease in the growth rate for electricity demand and the very effective global anti-nuclear lobby are all reasons for the decline in the nuclear power industry. Figure 4.4 shows quite graphically the decline in demand for nuclear capacity. There is considerable doubt that nuclear generation is the viable economic alternative that it was once thought to be.

Figures 4.5 through 4.8 detail the global reserves for non-replenishable fuels. Given these reserves it will be interesting to note the energy paths followed by the various nations in the coming years. For example, the nuclear programme in France, which will probably have some 13% too much generating capacity when completed by 1990 as compared with the more integrated fuel mix programmes of other European nations.

One of the factors pertinent to the development of energy programmes is the problem of lead time requirements. The lead time for alternate technology is significant since develoment may be perceived to be necessary to overcome crude oil shortages without undue economic hardship. For example, with reference to Figure 4.9, it would take upward of ten years to establish a commercial coal liquefaction plant including mine development. With known technology a commercial unit could be on line in, say, five or six years. However, a commercial unit is a very long way from a sufficiency of alternate fuel.

Globally the world is seeking alternate fuel sources of some 3% or less of current petroleum demand. This is essentially an offset to the demand growth. This represents about 1.5×10^6

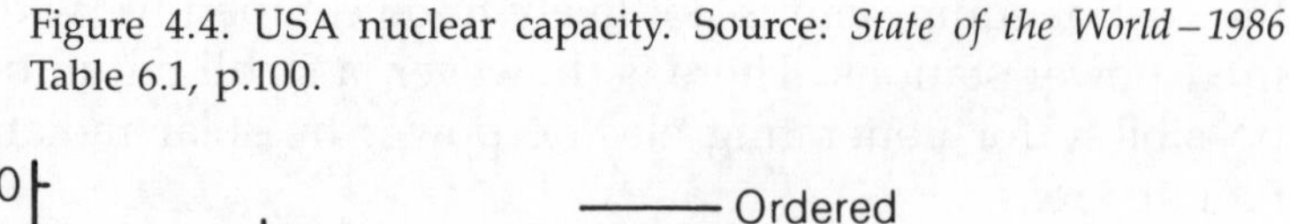

Figure 4.4. USA nuclear capacity. Source: *State of the World – 1986* Table 6.1, p.100.

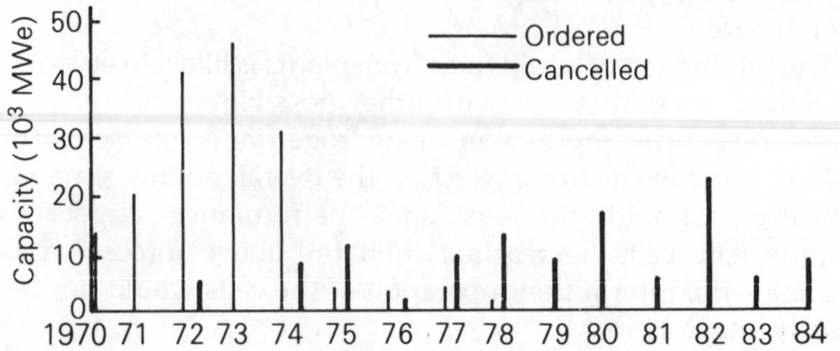

barrels per day capacity. Unless growth in energy demand is curbed, the later the alternatives become available, the greater the problem. The problem is one of inventing energy futures.

4.1 Inventing energy futures

In planning implementation strategy two types of innovation are to be distinguished: demand driven and technology driven. The former usually means the clearing away

Figure 4.5. Coal reserves (total in place). (Major Pacific region resources). Adapted from Miller (1981).

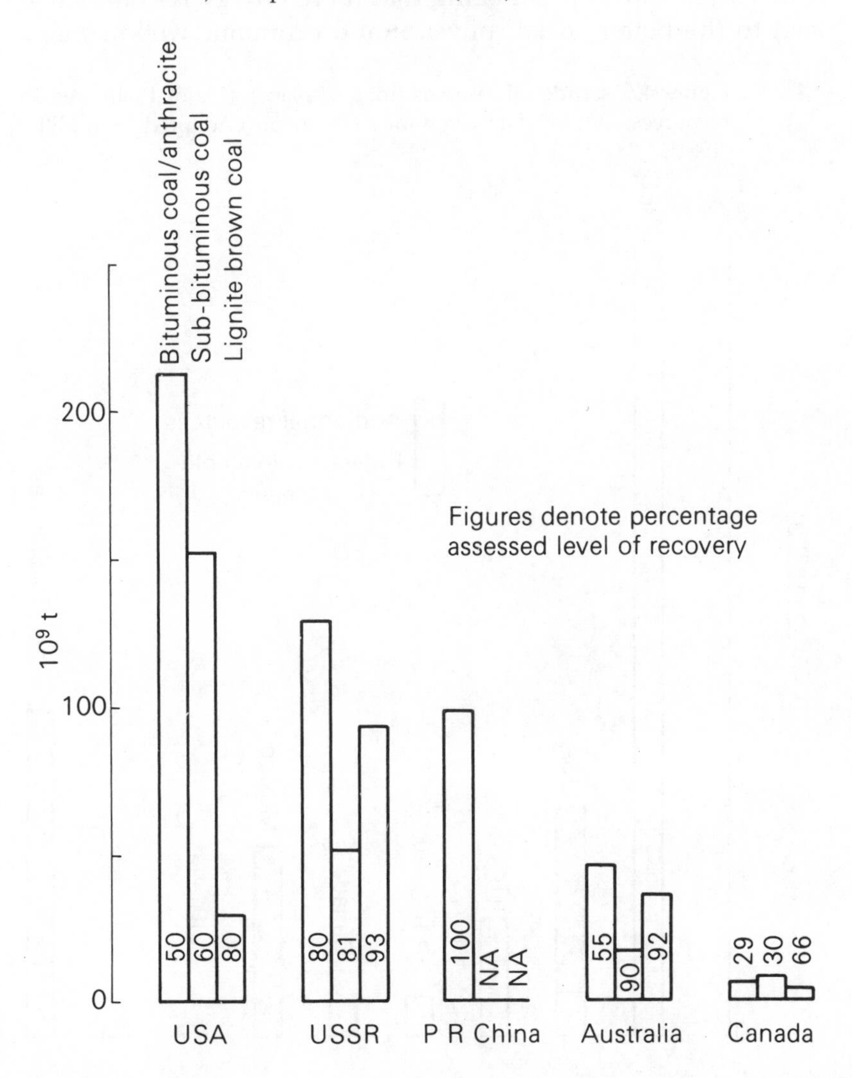

of impediments, the latter is usually seen in terms of demand creation.

Current governments see the energy problem as relating to the cost and availability of liquid fuels. Two aspects predominate policy thought: (i) short-term supply interruptions; (ii) long-term indigenous supply.

The process of development of energy policy is shown in Figure 4.10. Being essentially a political activity there are two basic views of energy policy formulation. The first, based on a growth concept, is geared to the development of new energy sources; the assumption being that more energy *per capita* will lead to the better social, physical and economic well-being of

Figure 4.6. Crude oil reserves (total in place). (Major Pacific region resources – OPEC data given for comparison.) Adapted from Miller (1981).

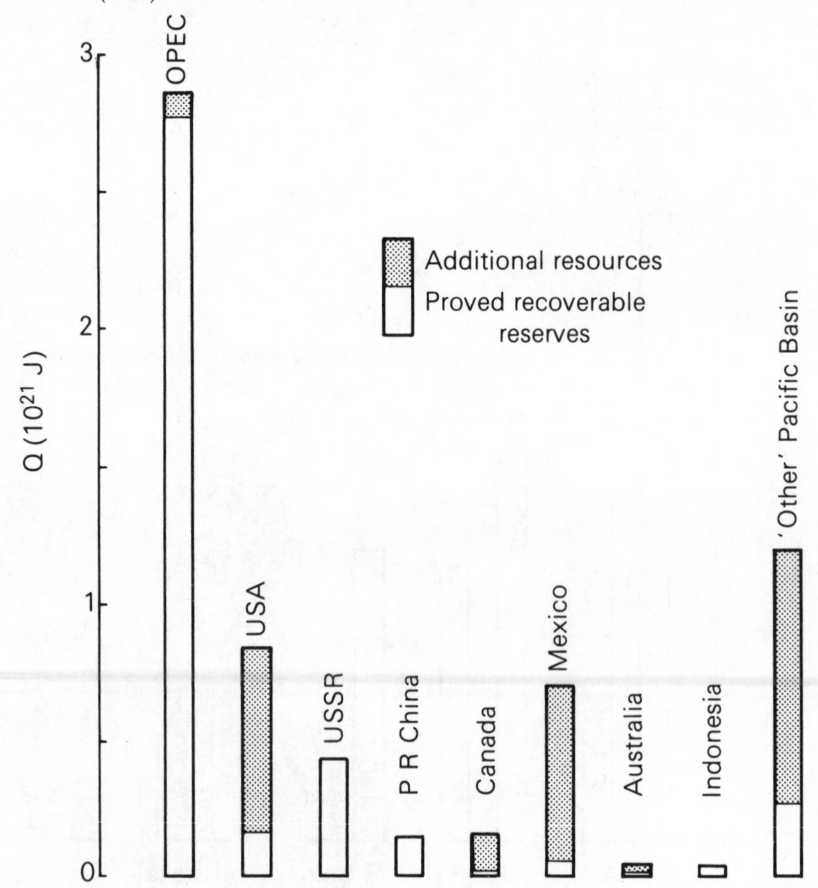

man. The latter, based on a steady-state economic scenario, requires a slowdown in industrial activity, which in turn necessitates a shift in socio-economic expectations and a change in life-style. The choice therefore is between a 'hard' energy path and a 'soft' energy path. The 'hard' path requires rapid expansion of centralised high technology to increase energy supply, especially electricity. By contrast, the 'soft' path requires

Figure 4.7. Natural gas reserves (total in place). (Major Pacific region resources.) Adapted from Miller (1981).

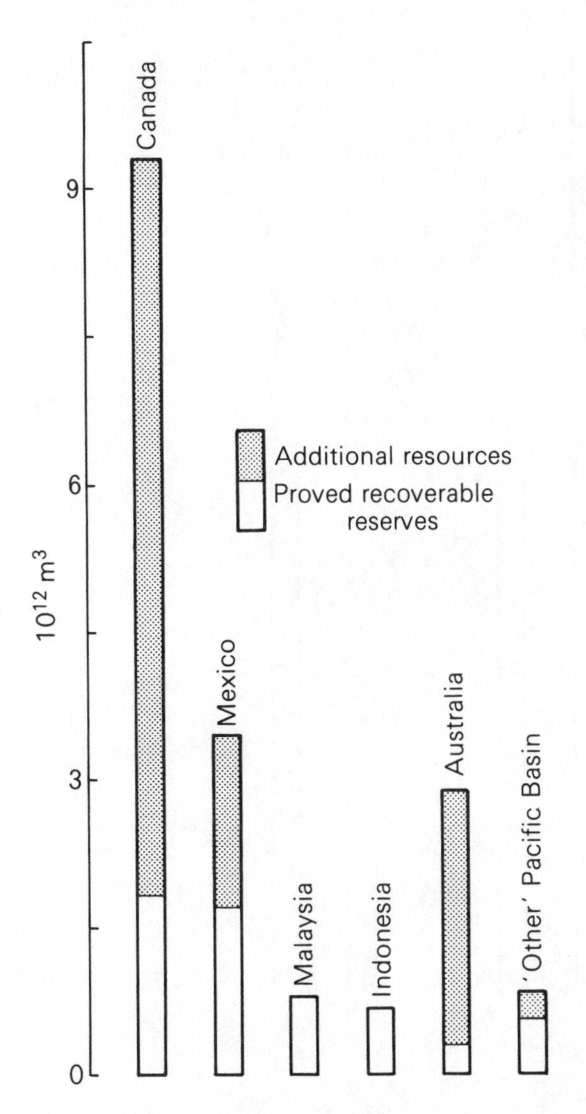

Figure 4.8. Uranium reserves (up to \$130/kg 1981 prices). (Major Pacific region resources.) Adapted from Miller (1981).

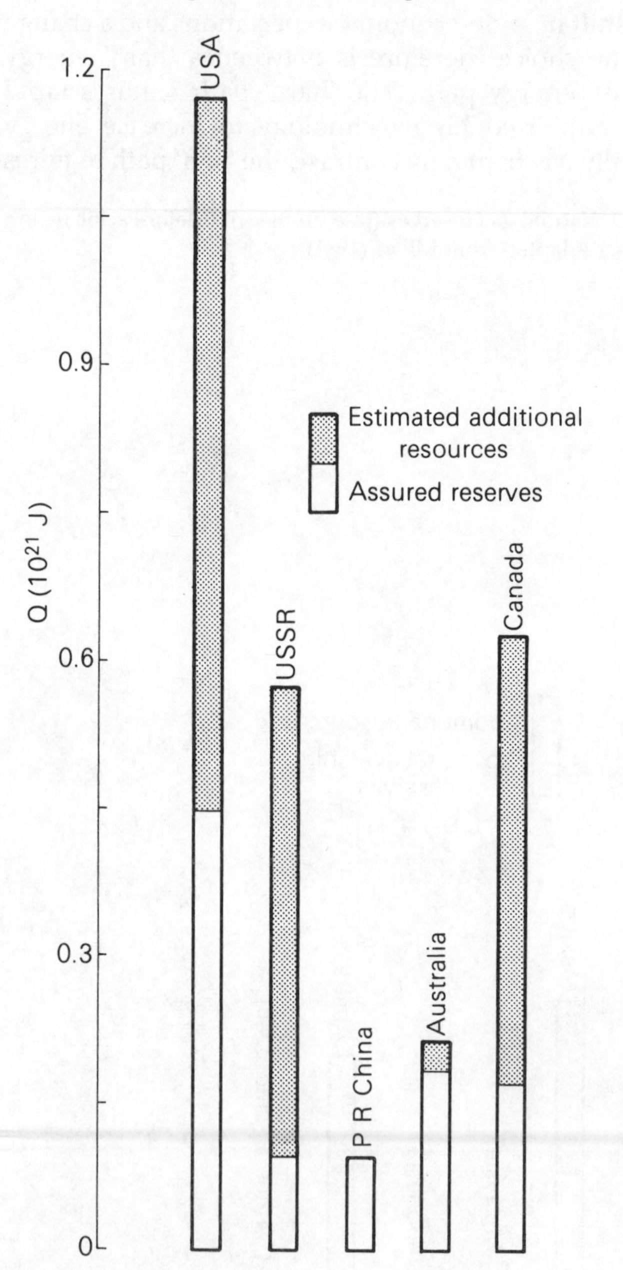

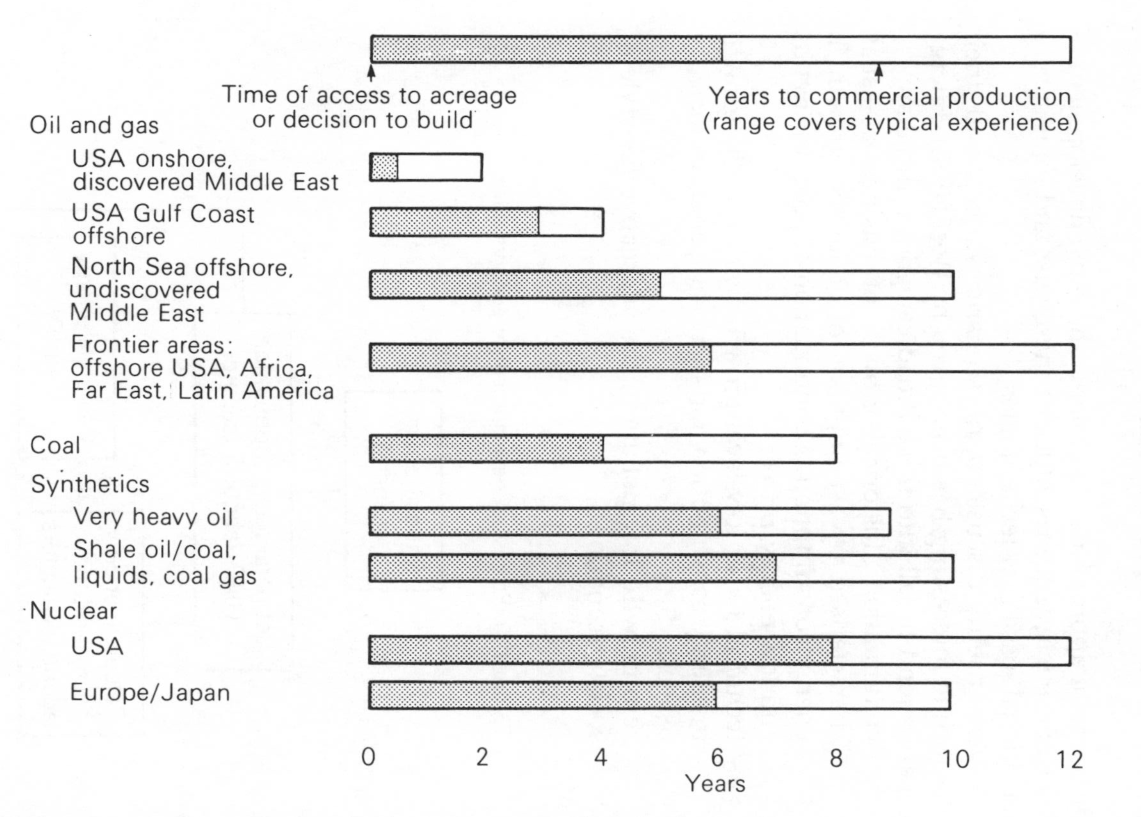

Figure 4.9. Energy supply lead times. Source: Shiels (1982).

the usage of appropriate technology, the rapid development of renewable energy sources, efficient energy use and some special transitional fossil fuel technologies.

A very useful discussion of scenario studies and their relevance to forecasting has been given by West (1978). Energy scenarios tend to fall into three broad categories; these are

- continuation, historical growth or business-as-usual, characterised by assumptions of about 6% growth in energy consumption, a continuation of post-war trends and low energy prices;
- technical fix or low-level growth, with a 2–4% growth in energy consumption and increasing energy efficiency usages;
- zero growth, low pollution, the alternative life-style or society scenario.

Figure 4.10. Process of development of energy policy.

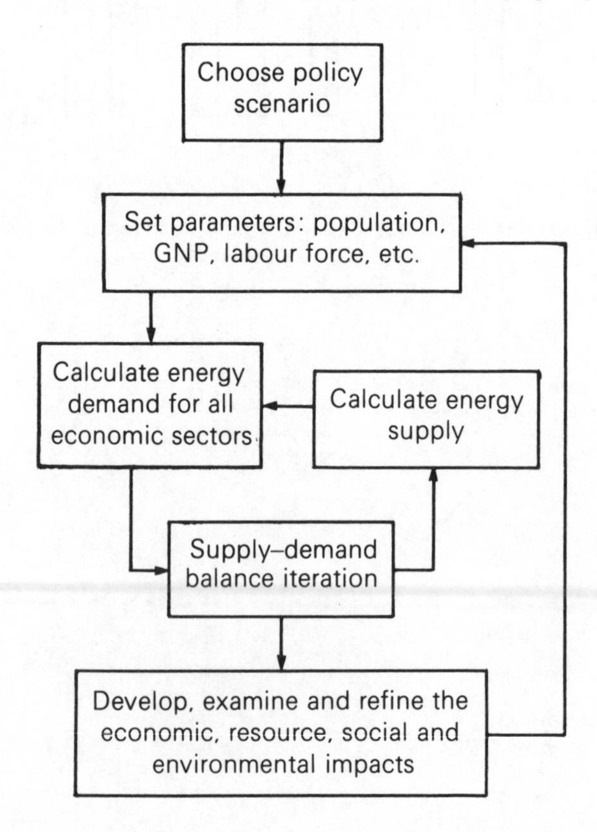

Generally each type of scenario is analysed in terms of parameters defining material wealth, environmental impact and resource depletion. The parameters themselves are related to the assumptions in input regarding population, life-style, progress of industrialisation and requisite energy demand. This type of analysis should not, although it often is, be regarded as a finite prediction. It is merely a method of comparison regarding the consequences of a line of action given certain defined inputs. A common denominator in most scenarios is the significance given to the obtaining and maintaining of full employment – more correctly, full paid employment.

The historical growth scenario is dependent upon the assumption that previous trends and aspirations will continue. Tradition is the keystone. The consequent energy policy is simply to direct the national effort into enlarging the energy supply to meet the expected demand. The idea of a limit is anathema to this scenario. There are many underlying assumptions in this type of analysis, but the major two are the belief that the external conditions for continued economic growth will continue, and that people and nations will persist in the quest for increasing material wealth.

In Australia this type of scenario necessitates the quest for more oil, or, failing that, the securing of increasing quantities of hydrocarbon imports, or oil substitutes. The emphasis is upon energy supply rather than consumption technology, and it is assumed that alternative energy forms will be readily available in the near future, i.e. AD 1990 to AD 2000.

There are inherent dangers in this type of philosophy, namely that the under-developed nations will not attempt to change the terms of world trade; nuclear power will be a reality; there will be a revival of the coal industry without causing additional pollution, and that maximum production is attainable from all available energy sources.

The technical fix scenario with its lower growth rates reflects a national or international recognition that energy should be utilised more efficiently. This type of growth stratagem reflects a greater security in terms of energy supply than the continuation model, with perhaps little disruption to traditional life styles. Energy conservation technologies and a desire for energy efficiency in conversion processes figure largely in this type of analysis. A possible plus for the technical fix concept is that through conservation a greater flexibility of energy supply mix is achieved. On the negative side it does rely heavily on the usage of increased and appropriate technology to obtain energy self-sufficiency; this could mean that substantial additional sources of energy could be required.

The zero growth concept places a great emphasis on the quality of the physical environment and the preservation of it. Oft times this type of model is quoted in the context that survival is more important than material affluence. The level of pollution consequent upon a course of action is the arbiter of desirability. Effective energy input per unit output is a prerequisite for action. The dominant theme is conservation and reduced pollution, hence there is a steady move toward renewable energy sources. The common misconception is that this type of scenario is concomitant with declining 'standards of living' and a preclusion of that all important desire for increased economic growth. The important directive is toward those economic activities which are not energy intensive.

The current government dictums on energy fit into the mold of the historical growth or continuation model. This line of action is not in the best interests of Australia. It is essential that a properly formulated policy on energy be developed for the well-being of the nation. It is to be deplored that our politicians and economists still support policies for ever-increasing industrial growth; policies which presumably are based on the assumption that this growth can continue indefinitely and will result in ever-higher levels of human well-being and evermore effective satisfaction of fundamental human needs.

5

Energy analysis and scenarios

Energy must be seen as part of the total world economic system; Australian usage cannot be divorced from the rest of the world, nor can energy be separated from other aspects of society and its organisation. All transactions in industrialised societies are energy dependent, but the very complexity of the societal structure often precludes the definition of energy as an explicit variable. The flow of energy through a system, with the aid of energy evaluation techniques, can be quantified so that the primary energy inputs into a production system can be identified. In this type of analysis, the law of energy conservation is assumed to be valid, i.e.

$$E = \Sigma\, q \cdot e$$

where E = total energy consumed by the system; q = amount of commodity produced in the system; e = gross energy requirement for q. Simple summation is then made of all commodities produced in the system. There are four basic methodologies of energy analysis: input–output analysis, statistical analysis, process analysis, and interpretation by means of the 'second law of thermodynamics efficiency analysis' (SLTE). (Finney, 1976).

The statistical services of most governments provide input–output listings showing the sectoral inter-relationships of the national economy. From these tables it is possible to define the amount that each sector consumes in respect of the output of all others: for example, the amount of cement, timber and other materials consumed by the building industry. If the initial direct energy input into any sector is known, then by iterative calculation the energy intensities can be defined. The energy intensity is the ratio between the total energy input into a sector and the financial value of the sectoral output. Various errors do arise in this form of analysis: discount and preferential pricing, varying process technologies for a product, and statistical readjustments.

Statistical methodology relies on published statistical data for a comparison of technologies and technological process in defining output. This type of analysis is not very popular because, firstly, rarely is statistical data collected or presented with energy analysis in mind; secondly, rarely are all inputs to a given sector available; and, lastly, counting is difficult. Finney (1976) gives examples of this type of analysis in defining the energy statistics of Tasmania.

One of the most popular forms of energy analysis is called process analysis. All the physical and chemical transformations occurring in a particular production process are identified, and energy values assigned to each input. Pearson (1977) has indicated that three types of energy requirement can be distinguished; namely,

(*a*) process energy requirement (PER), which is defined as the total fuel energy supplied to drive all process stages;

(*b*) gross energy requirement (GER), the amount of energy source sequestered by the process of making goods and services;

(*c*) net energy requirement (NER), defined as the GER minus the gross heat of combustion of the products of the process.

There are, however, serious problems associated with process analysis. The first is the determination of the boundary of the production system and the definition of the level of analysis. For example, boundary problems are easily defined for the mining process, and cut-offs may operate at the mine head, concentrating or smelting stages. In terms of the mineral industry, the final cut-off point is not so easily definable: should it include the car body, or merely the sheet steel? The level of analytical definition is more often than not governed by restrictions imposed by the available data.

Further problems occur in respect of allocating energy requirements when more than one distinct commodity is being generated by a particular process. The solution given by IFIAS

(International Federation of Institutes for Advanced Study, 1975) is to partition by defined physical parameters. Added complications arise when a particular commodity may itself form an input to its own production process as, for example, when blast furnace charges include a proportion of scrap iron produced in earlier operating cycles. Final energy requirements can then only be defined by iterative calculation.

The SLTE was pioneered by Berry and Fels (1973). In this technique the minimum quantity of energy required to perform a given task is calculated, and used as input. Simple summation then provides the energy cost for the output product. SLTE analysis therefore outlines the theoretical, or maximum efficiency, energy requirements for a particular process and provides an efficiency measure for industrial output, which has direct application to the mining and minerals industry. The theoretical energy 'cost' can be compared with the actual energy requirements so as not only to define the current efficiency of the process, but to give indication of future energy requirements, and to define the process areas where technological improvements might result in energy savings. This type of energy analysis is compatible with process analysis, and also suffers the same problems in usage.

A convenient way to measure energy efficiencies is in terms of productivity. Generally this is done in terms of economic gain, the value added per unit of energy. Commoner (1978) recognised three basic productivities that should be considered. These are the energy productivity, the capital productivity and the labour productivity. Energy productivity relates to the efficiency of energy conversion into value added, *viz.* dollars per energy unit. Capital productivity likewise measures the efficiency of capital investment conversion, and labour productivity is the efficiency of labour conversion to dollar added value. There are basic links between energy, the economic system and the environment. In many ways energy has created a cultural norm in society and imbalances between the living standards of the developing, developed and impoverished regions.

Technological development and industrialisation has become extremely energy intensive. About 99% of the driving force in modern industry is geochemically stored and transformed energy.

Conservation is of course an energy source that produces no waste, but neither does it produce export dollars, but it does give 'value added'. In essence, though, conservation is an ethic implying the long-term management of natural resources for the benefit of mankind. What is required therefore in all of our policy formulations for energy is a natural resource ethic: a code of conduct, a set of moral principles defining the art of living in and with our environment.

The problems to be overcome in the definition of our energy scenarios are shown in Table 5.1 and Figure 5.1. The problems confronting us for the future are not insurmountable, but global societies of all political persuasions will have to recognise that overdevelopment is as severe a handicap to human well-being as underdevelopment. Scenarios should not be chosen purely for the sake of non-attainable goals, namely ever-increasing growth and eternal progress. These are nonsensical myths.

Figure 5.1. The quality of life. EI = Function of population (P), *per-capita* resource consumption (C) and environmental deterioration (D) of life support system = $f(P \cdot C \cdot D \cdot)$. Adapted from Ophuls (1977, p. 133).

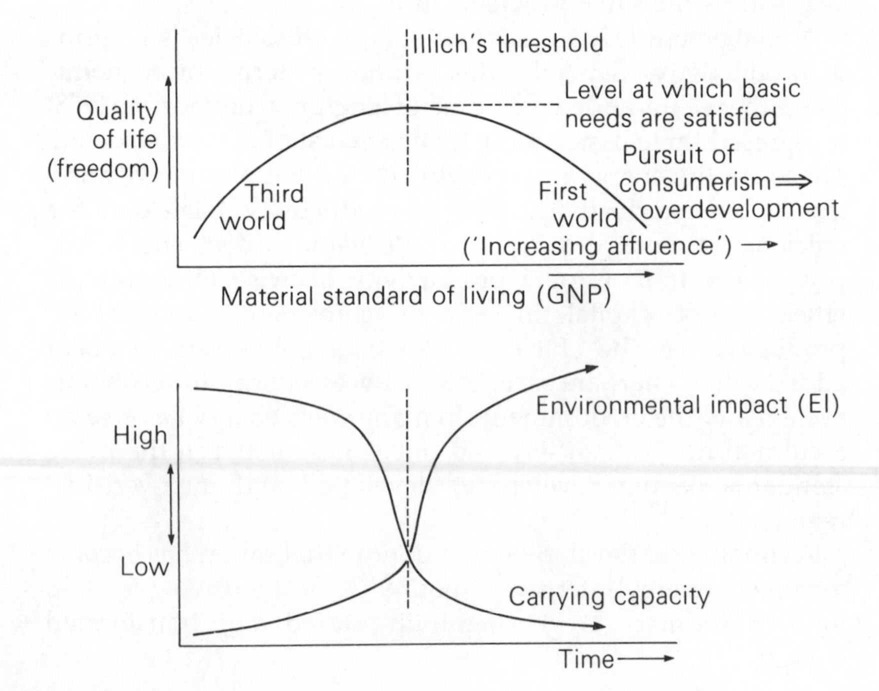

5.1 Energy scenarios

Figure 5.2 shows the energy flows defined in an industrial economy. This type of model is a reflection of the energy flows through most industrialised nations although the actual type of energy fuel used will naturally vary. The actual volume of energy flow, the energy requirement, will be dependent upon the type of scenario defined in a country's energy policy. This policy may be either explicitly stated as documented policy and/or implicit and defined solely by the scenarios developed.

In the preceding section we briefly described the three broad scenarios commonly followed. Here we wish to expand the analyses. The three scenarios are usually defined as

- 'Continuation', 'historical growth', or 'business-as-usual'; with a high energy growth of 5–6% *per annum* based on post-war trends and low energy prices.
- 'Technical fix' or 'Limited growth'; with a moderate energy growth of 2–4% *per annum* based on current energy prices and improved energy efficiency.
- 'Zero' or 'low growth/low pollution'; with energy growth rates that move fairly rapidly towards zero or even negative rates of growth, as a result of an explicit desire to avoid the apparent ill effects of higher rates of consumption.

Each scenario has a feature which is dominant in the analysis of energy demand and supply. The three features usually

Table 5.1 *Problems confronting the future*
Source: Higgins (1978)

Special Societal Problems	Manifestations of societal problems
1 Population growth	1 Bureaucratic insensitivity
2 Inequitable resource distribution	2 Self indulgent consumerism
3 Incompatible political systems	3 Sociopathic greed
4 Environmental deterioration	4 Fallacy of market forces
5 Abuse of nuclear power	5 Environmental despoilation
6 Failure of science/technology to provide Utopia	6 Irrational epidemics
	7 Consumer-oriented education
	8 Structural employment

considered are material wealth, concern for the environment and non-depletion of resources. These give rise to differing assumptions concerning population, life-style, type/level of industrialisation and energy supply. A common and important feature to all scenarios is that the analyses are based on the assumption of full employment.

It should be remembered, however, that scenarios are not long-term predictions. They merely aid in the comparison of the consequences of different energy choices. In reality, there are an infinite number of energy strategies available. The purpose is to spotlight a number of possibilities in order to: clarify the implications of different rates of energy growth; determine resources needed to make each strategy work; and to ascertain the effects it will have on the economy, the environment, our life-styles and the cost of energy.

The historical growth scenario

This model posits that there will be no change in the historic patterns of energy usage and neither will any policies be implemented to effect any change. The trends and aspirations currently functioning in the economic system are assumed to continue. Energy policy as such is directed toward enlarging the energy supply so as to meet the increasing demand.

Businesses are assumed to continue to seek growth to guarantee security and to increase disposable income. The major political and economic institutions would continue to exist and exert pressures in their traditional roles. A most important assumption underlying the historical growth philosophy is that the external conditions for continued economic growth continue to exist. Apart from the obvious political assumption that people

Figure 5.2. Energy flows

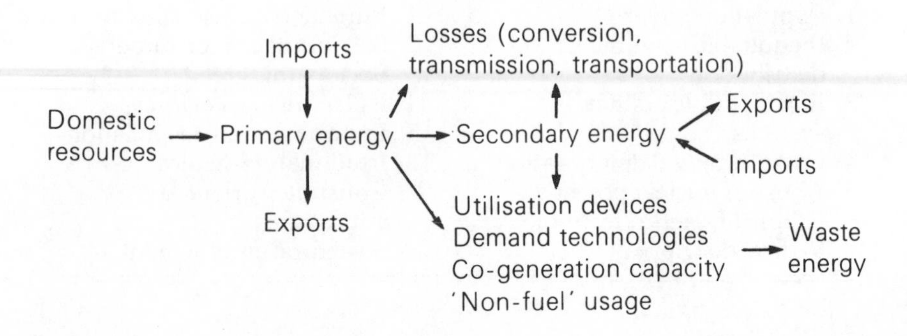

will continue to want increased wealth, another major assumption is that the rest of the world will also adopt an historical growth philosophy.

In Australia, this scenario necessitates substantial oil imports causing problems with balance of payments. Alternatives to large oil import bills include reversing the roles of electric power and oil and gas, the development of supplemental uses of coal and oil shales, and the conversion of these resources to liquid and gaseous fuels. There are problems with alternative fuel development programmes, and synthetic fuels are not expected to play a significant role in energy supply before AD 2000.

A strong energy research and development programme is essential to support this scenario, and emphasis should be on energy supply rather than consumption technology. For the success of this scenario it is assumed that each new energy source will be developed and in production before the year 2000 (e.g. oil shales, coal conversion).

The historical growth scenario conforms to present social expectations of the Western world, and presumes that

- all of a nation's trading partners will also adopt an historical growth scenario;
- underdeveloped nations will not form cartels or otherwise change the terms of world trade; and
- maximum uninterrupted production from all available energy resources will continue.

The continuation of past growth patterns is contingent upon the maintenance of relatively low energy prices. Such demand growth is therefore only likely to be realised if energy prices to the consumer are set by subsidy probably at levels below true production costs. Continued growth at rates approaching those of the past will therefore only be possible with massive government commitment. The recent energy policy programmes of most of the Western nations would seem to indicate that no government is seriously considering the implementation of this type of scenario as a credible energy future.

The technical fix scenario

In terms of the mix of goods and services this scenario would differ little from the historical growth scenario. The rate of economic growth would, however, be much slower. It would seem that this type of scenario offers a greater future security than the historical growth scenario without seriously upsetting the social expectations or life-styles of the population.

Such a scenario reflects a conscious national and/or international effort to use energy more efficiently, with energy consumption growth rates of between 1.5 and 2.5% annually. In this option, problems of finding future energy supplies are significantly reduced by reducing energy demand and waste. It is presumed that reductions in fuel demand can be accomplished by measures which do not significantly alter present life-styles or decrease standards of living. Essentially, opting for this type of strategy indicates the belief that an historical growth scenario will fail, and that it is preferable to accept a planned reduction in the growth of energy consumption.

Energy savings fall into two categories. The first involves direct energy savings, resulting from the application of energy conservation technologies at the point of energy use. Increased thermal insulation, heat pumps and improved automobile efficiency and fuel economy are examples. The second involves indirect energy savings that can be made in the energy processing sector.

This scenario assumes that, in the future, energy users will become more aware of energy costs and the need for energy conservation. This is especially relevant to the energy intensive industrial sector. If there are no effective national energy policies that supplement public awareness, then it is likely that only significantly higher energy prices would achieve the low growth rates required in a technical fix energy future.

A basic advantage of the technical fix scenario is that, through energy conservation, flexibility in determining an appropriate energy supply mix is achieved. Even with the low energy growth rate in this scenario, substantial additional energy resources are required. However, it is possible to forego development of some major energy sources with this option. Thus, energy self-sufficiency and environmental protection are two basic supply strategies for the technical fix scenario.

The zero or low growth scenario

This type of scenario is based on the value judgement that survival is more important than increased affluence. It is a scenario that places great value on the quality of the physical environment and therefore imposes restrictions upon the goal of high material consuming standards of living.

One of the principal ways in which concern for the environment is manifested in this scenario is to restrict levels of production in those activities which are seen to be major causes of pollution. The restrictions placed on these activities reflect the overall concern to reduce energy demand, since all sources of energy production have detrimental environmental effects to a greater or lesser extent. Effective energy input per unit output is reduced by varying amounts in different sectors, depending on conservation potential and the energy requirements of cleaning up pollution. Measures taken to improve the environment will cause some reduction in the output of material goods, and subsequently there is a low GNP growth rate.

There are other reasons why this scenario is seriously considered. It has been argued that factors such as 'loss of community', 'lack of job satisfaction' and 'loss of amenity' are more important than benefits thought to accrue from increased material possessions. This scenario does not require an austerity programme, nor does it preclude economic growth.

Within industry, manufacturing would continue to grow and reach a higher level than today, both on an absolute and *per-capita* basis. Since measures such as an energy sales tax would raise the price of energy intensive materials relative to other costs, less material would be used in products and growth in output would decline. Individual industries would be expected to grow more or less slowly, depending on whether they were energy intensive and upon the response of buyers to the prices of their products.

The energy supplies required for zero energy growth are not simply scaled down versions of the supply schedules for higher growth scenarios. It is anticipated that, initially, conventional fossil fuels would continue to be the major energy suppliers, but there would be a growth in alternative energy forms, such as gravitational, solar and geothermal, as they become technically and economically feasible.

Figure 5.3 outlines the varying consequences in energy terms of the choice of particular scenarios. One particular problem of zero growth or low growth scenarios is the choice of energy level (total or *per capita*) at which energy consumption should plateau. For example, with respect to Figure 5.3 zero growth scenarios for the USA show a level in consumption at about 106 EJ *per annum*. The EEC figure is slightly more than half this value at 56.7 EJ *per annum*. On a *per-capita* basis this represents a USA consumption of about 390 GJ *per annum*, whereas for the EEC the figure is only 158 GJ *per annum*. Zero growth scenarios tend to 'freeze' a society at a particular level of material well-being. The problem is that most nations appear to want the USA level of consumption irrespective of the social considerations.

The three scenarios discussed above are just a few of the many possibilities. Other scenarios, such as no nuclear, all nuclear, increased energy efficiency, reduced energy imports, all renewable energy sources, can also be developed. However, whatever scenario is used to guide a nation, the policy elements of that scenario must be quickly implemented, and be flexible enough to allow for redirection.

The timing for achieving zero energy growth is equally vague

Figure 5.3. Schematic representation of scenarios – USA and EEC. (*a*) High growth. (*b*) Technical fix. (*c*) Zero/low growth.

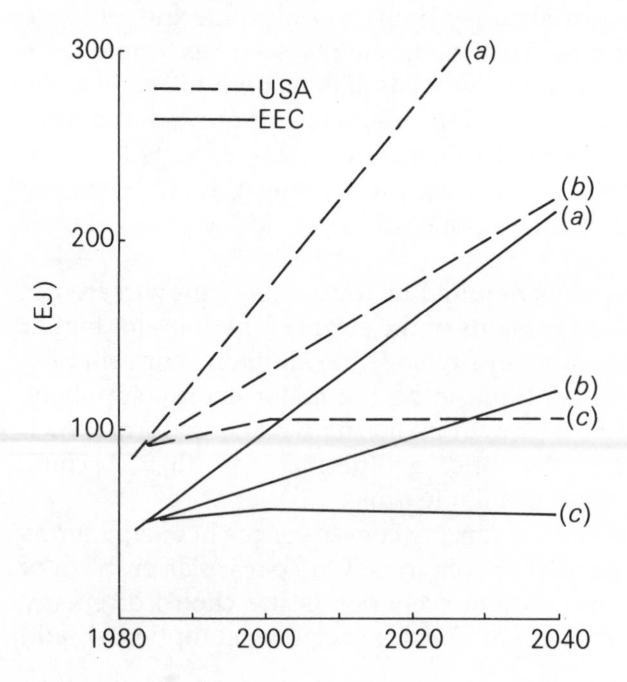

and uncertain. On thermodynamic grounds zero energy growth is possible immediately with only efficiency increases. However, the necessary institutional changes to affect it are unlikely to occur quickly. Moreover, the industrial and commercial conversions could not take place rapidly. Whilst a zero or low energy growth policy is the only long-term option open to the world, the social, economic, technical and political conditions prevailing in most Western nations are such that the immediate adoption of this strategy is unlikely.

This conclusion emphasises one of the serious drawbacks implicit in the scenario approach to formulating policy. Throughout the evaluation of any scenario the same set of assumptions are applied for long periods of time. This is essential if one is to see the implications of any one policy carried out to its end; however, it tells nothing of all the options which arise if the policy is changed or modified at some point in the future. Thus, as time proceeds, there is an increasing degree of uncertainty associated with the scenario.

Despite the attractiveness of the zero energy growth option, there are numerous practical problems involved in its immediate adoption; thus, one opts for the compromise, the technical fix strategy. The major danger with such an approach is that it is unlikely to arouse the level of emotional commitment the other scenarios evoke from their respective advocates and/or opponents.

Part 2 The mineral resource system

6

Mineral resources

Fundamentally all mineral resources, excluding the fossil fuels which have finite limits, exist within the Earth's crust in quantities far exceeding man's present or foreseeable future needs – even with a growing population and increasing individual demands. The problem of the supply of mineral products is a function of the economics and technology that provide an assured source at a proper price. Supply is principally a function of energy and capital input to affect metal or mineral output, within the limits of acceptable environmental effects, and offering an attractive return on the investment. There appears to be a growing tendency to give contract preference and 'bonus' prices to sources that offer 'security of supply', as raw material costs become a smaller proportion of total costs when compared with capital and labour. When the cost of any mineral product rises too high or its production is uncertain, substitutions tend to take place.

Within the limits of a given set of economic and technological constraints, each individual deposit must sooner or later be exhausted and therefore new discoveries must be made. This being so, there should always be national concern whenever reserves of any mineral commodity, that is a major contributor to economic health or national security, reaches a level of less than 30 years supply; finding and developing new deposits is a slow process. There is therefore a constant concern amongst those engaged in the mineral extractive industries in respect of the 'discoverability' of new ore reserves that are economic at a specified period of time.

In the early stages of mineral resource development in Australia, deposits were at or near the surface and many high-grade deposits were discovered by accident. More recently, the discovered ore deposits have generally been of lower grade and/or less accessible. In this context they require more knowledge, energy and capital to find, evaluate and develop. This has led to very substantial increases in investment costs. In Australia over the past decade, it has cost an average of 5% of the total contained metal value to discover a new mineral deposit and an additional 20% of the total contained metal value to develop the deposit to the production stage. Preproduction time ranges from five to in excess of ten years, providing capital finance is available.

Technically a mineral is definable as a naturally occurring crystalline solid of inorganic origin and possessing a more or less defined chemical composition. Petroleum, gas and tar sands and such are commonly included in the above definition, although they do not actually conform to it. The term mineral is often somewhat loosely used to describe any substance that is useful to man that is extracted from the Earth. Fossil fuels, coal, gas, etc., which are thought to be produced by organic processes, are also often termed raw materials, a term which includes soils, water and minerals. All are essentially non-replenishable with the exception of water. In this sense there are some 2000 minerals available to man. Of these, though, only about 100 are generally considered to be useful.

The inorganic minerals are sub-divided into the metallic and non-metallic (industrial) minerals. The metallics, or metals, include such minerals as copper, lead, zinc, etc. They are commonly termed base metals, and include most metals basic to the heavy industries. The non-metallics include a very wide and diverse group of minerals – clays, gypsum, phosphate, sulphur and the precious/semi-precious 'stones'. Sands and gravels are also included within this sub-group.

Now although for many minerals, excluding the fossil fuels, there are abundant resources, the reserves are limited. A resource estimate is nothing other than an estimate of what is contained within the Earth's crust. A reserve is an estimate of what actually is economically and/or thermodynamically extractable. A cut-off grade is the lower limit of mineral concentration that can be extracted under given economic or thermodynamic consideration. The relationship between a resource and reserve is given in Figure 6.1. The actual definition of what is termed a reserve is shown in Figure 6.2. A reserve is termed economic if it is extractable at a profit. Measured reserves are those that have been completely outlined *in situ*. Indicated (probable) reserves are those which have been partially outlined *in situ* – say two or three sides of the mineral block are visible and have been measured. Inferred (possible) reserves are those that have been assessed as being available on geological evidence.

Within any time framework there are various factors or parameters defining the reserve – resource boundary conditions. The boundary conditions are the natural limits, the production limits, the market limits and the legislative procedures and constraints. The relationship between the legislation constraints and market limits define the degree of rationing of supply in times of deficiency and stock-piling in times of surplus supply. The relationship between the natural and production limits is a function of thermodynamics manifest as the method of extraction and milling. Within these boundaries the calculus is defined by particular taxation policies, land availability, tenure, costs of production, population, economic controls and the degree of recycling and substitution.

The boundary between what constitutes a reserve and a resource is therefore a complex time varying phenomenon and only definable under fairly rigid conditions.

6.1 Mineral systems

Figure 6.3 shows the Australian mineral industry as a nested hierarchy of systems defined by a global system and the inter-related national systems. This policy environment is typical of most of the industrialised Western nations. This figure depicts the whole mining industry as a system within a much larger world system which embraces the whole international social, economic, technological and political environments.

The international environment includes both government interaction and the activities of many mining multi-national corporations, which by their size and degree of vertical integration are capable of controlling both the supply and

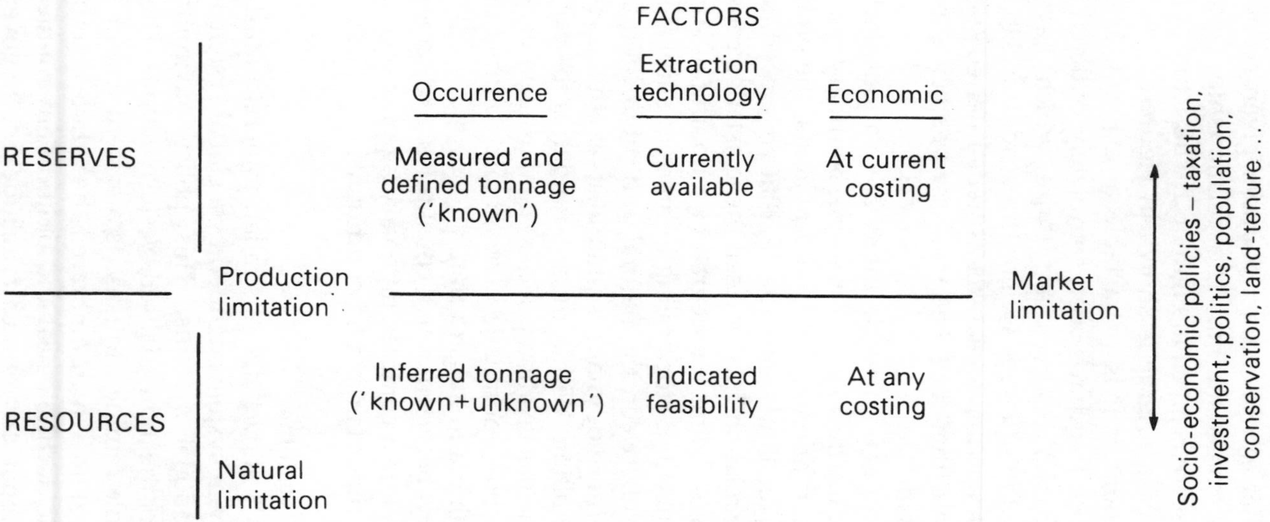

	FACTORS		
	Occurrence	Extraction technology	Economic
RESERVES	Measured and defined tonnage ('known')	Currently available	At current costing
RESOURCES	Inferred tonnage ('known+unknown')	Indicated feasibility	At any costing

Production limitation

Market limitation

Natural limitation

Socio-economic policies – taxation, investment, politics, population, conservation, land-tenure . . .

Figure 6.1. The characteristics defining reserves, resources and resource base. Adapted from Lacy (1983).

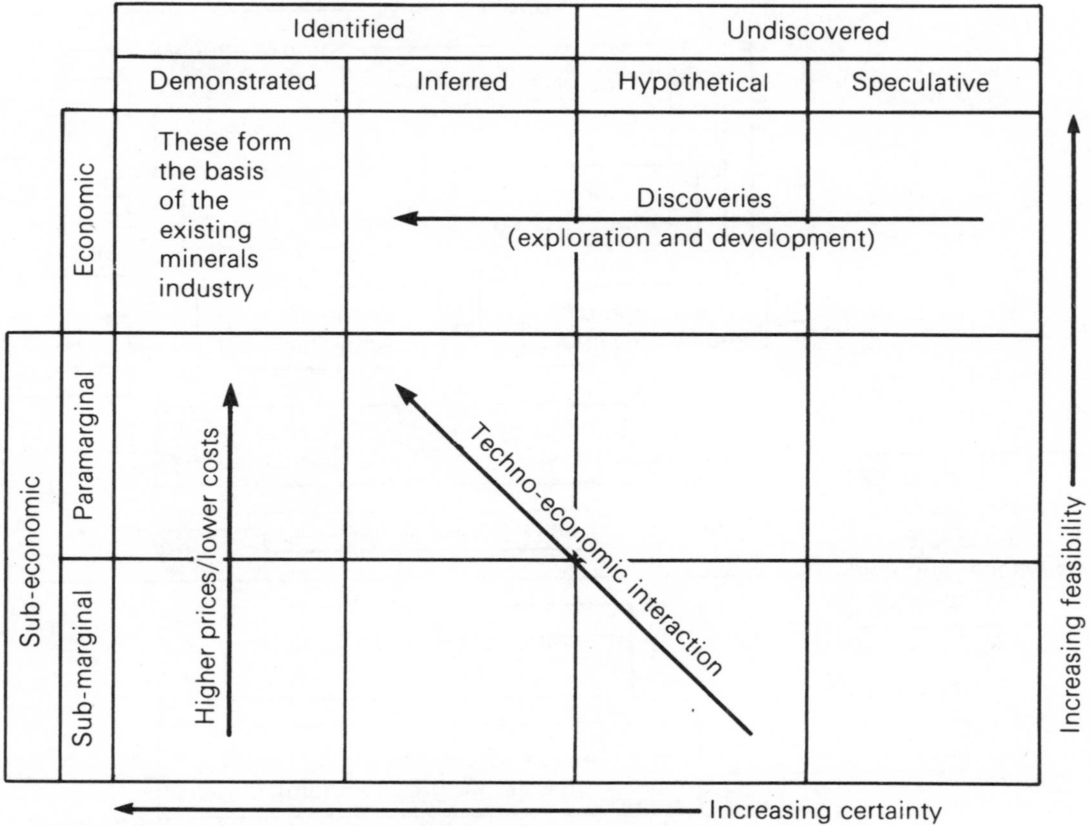

		Identified		Undiscovered	
		Demonstrated	Inferred	Hypothetical	Speculative
	Economic	These form the basis of the existing minerals industry	Discoveries (exploration and development)		
Sub-economic	Paramarginal	Higher prices/lower costs	Techno-economic interaction		
	Sub-marginal				

Increasing feasibility

Increasing certainty

Figure 6.2. The McKelvey box. Source: Dick & Mardon (1979).

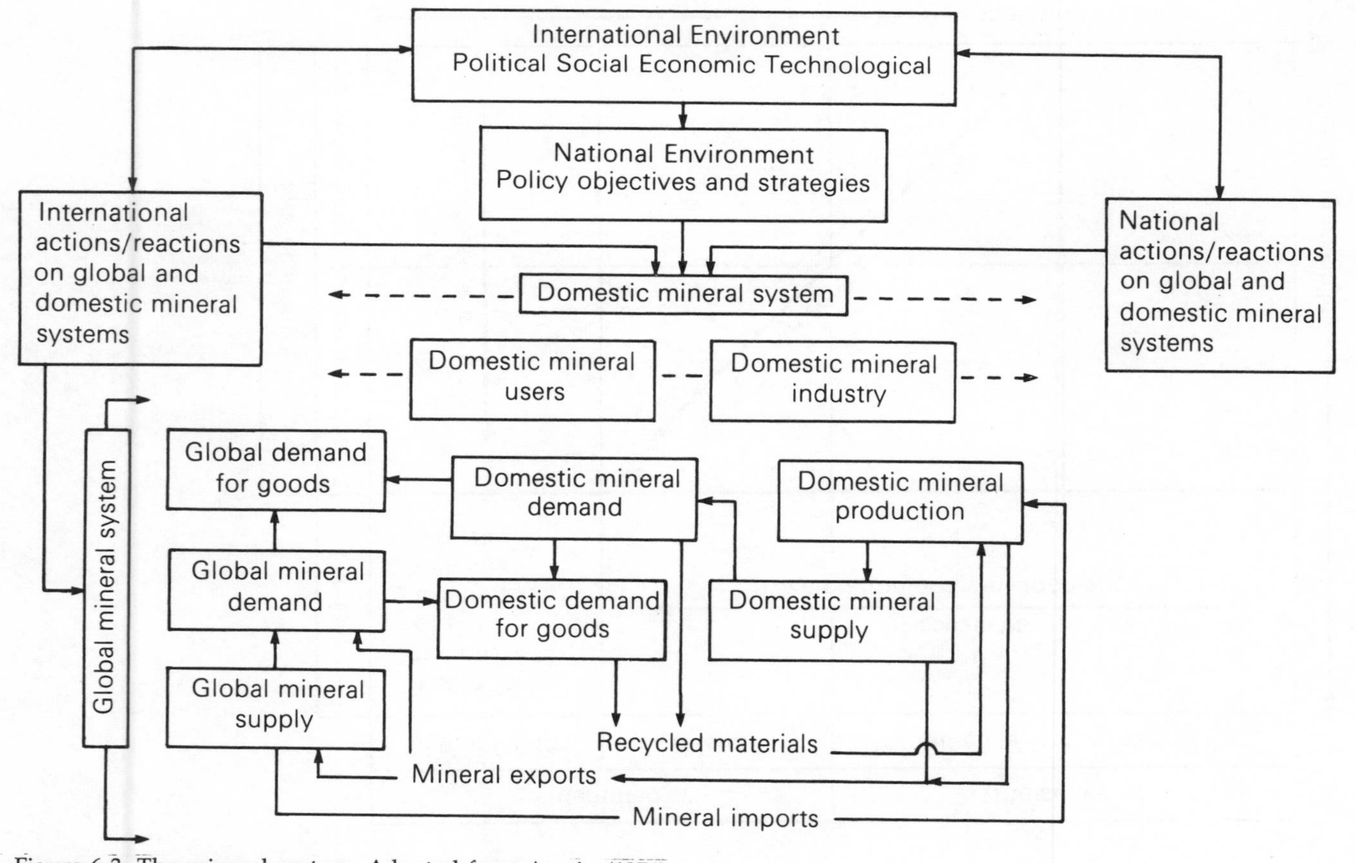

Figure 6.3. The mineral system. Adapted from Austin (1974).

demand. The aluminium industry is an example of this phenomenon wherein six companies virtually control the 'free-world' global trade.

The global demand for minerals creates a market for which the national supplier can compete. The domestic market is essentially a closed loop often maintained by tariff barriers or trade sanctions. The surplus is then sold on the market either by spot sales or in fulfilment of contractual obligations. Problems occur when the systems which inter-relate are moving (i.e. producing) at widely disparate rates. A recent example of this is the copper industry wherein many third world producers marketed their product at below cost simply to maintain a cash inflow and service their creditors.

The recent phenomenon of the multi-national corporation in the mining industry has meant that the national system boundary has become blurred with time. The multi-national corporation owes no national allegiance and is purely an exploiter of mineral resources.

The ability of any country to compete in the international trading environment is affected by a number of factors. The first determinant is price. Prices of most major minerals are determined by movements in the international levels of supply and demand. Various international commodity associations, such as the now defunct Tin Council, attempt to control the price so as to avoid extreme fluctuations. In some cases the mineral price range, especially for non-ferrous metals, can vary by as much as 100% in any one year. There are very few countries which have a monopoly situation on any one commodity. Rare exceptions do exist, as, for example, platinum, but these are due more to political influence rather than geographic distribution.

For most producers the mineral industry represents a highly competitive market and for most, and especially Australia which is distant from traditional markets, it is imperative to contain and minimise production costs. The sheer size of the Australian landmass and the general remoteness of most deposits adversely affects the cost structure at all stages of exploration, development and production. In addition, the labour costs in Australia are amongst the highest in the world, and the cost of constructing a mine in Australia is at least 20% greater than the cost of an equivalent project in Canada (*Mining Review*, May 1984, p.17).

The last factor is the level of domestic inflation. Any sustained relatively high level of inflation seriously affects the industry's competitive position.

6.2 Mineral resource scenarios

The 20th century has seen an explosion of technological development, especially within the electrical, electronic and transport industries. In the last few decades many new metals have come into prominence. Through rapid expansion of application, resulting from consumer demand, possibly the resultant of aggressive advertising, these have become established as basic components of modern living. In almost all aspects of our lives we rely on minerals. Modern man's desire for these products to improve individual living standards, and the community's need for metals and their products, have meant that the mining industry has had to find and develop more and more mineral deposits. This technological advance has provided the main impetus for the increased development of Australia's mineral resources.

Figure 6.4 shows the historic growth pattern of global consumption patterns in iron ore, copper and lead. The problem with respect to mineral reserves and resources is not so much pertaining to mineral exhaustion but the stabilisation of levels of consumption that are globally realistically attainable. For example, for a global population of some 6.5 billion in the year

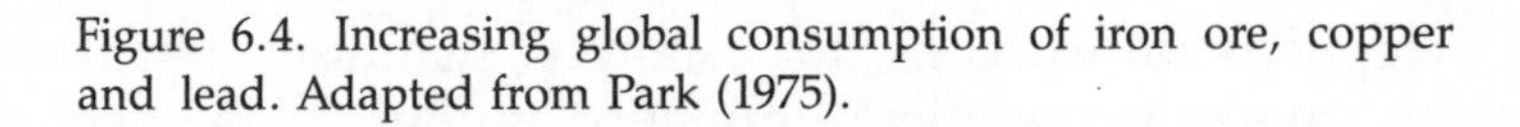
Figure 6.4. Increasing global consumption of iron ore, copper and lead. Adapted from Park (1975).

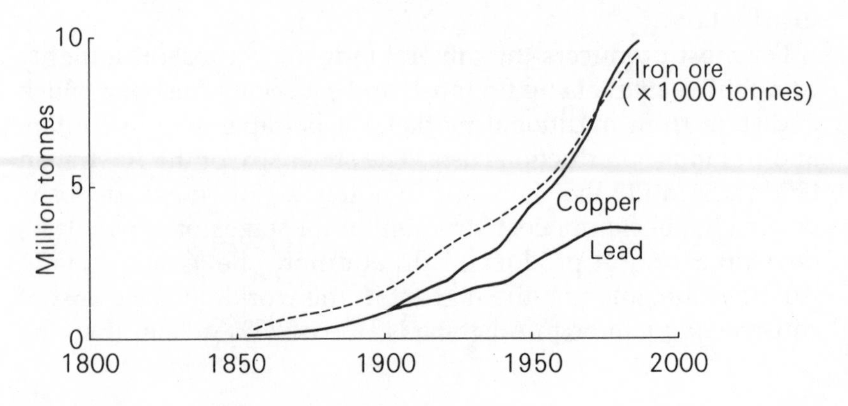

AD 2000, to raise the global consumption level to a comparable level to the USA would require increasing the production of iron ore some 12 times, that of copper about 11 times, and lead some 16 times. These levels are simply not realistically attainable.

Supplies of what are termed the geochemically scarce elements such as copper, lead and zinc are apparently proportional to their crustal abundances. At the present time we are using these metals at rates that are not proportional to their abundance. For example, we are using copper and lead at a rate about 40 times faster than that for iron, a geochemically abundant element. In contrast to the geochemically scarce metals, all the geochemically abundant metals are under-used. However, as the grade of ore declines, the energy input, which is an irreducible input per unit mass of metal recovered, rises steadily. This then leads to the third and final problem, the energy cost of mining and mineral processing. For example, the energy required to produce 1 kg of copper goes from 101 to 5170 GJ as the ore grade drops from 0.7 to 0.01%, Barnett (1979). This input is irreducible no matter what technology is used. The problem is to find a 'cheap' source of energy.

The energy cost of mining and processing is related to four main criteria: the grade of the ore, the quantity of consumables required, the complexity of the processing and, lastly, the scale of operations. From the mining perspective the most important criterion is the grade of the ore. As ore grades decline from 21 to 2.5%, the energy requirement increases ten-fold. A further decrease to 0.7% involves a further three-fold energy increase.

During the greater part of the industrial era, however, minerals have been getting cheaper. A study by the Paley Commission in the USA in 1952 stated that with the exception of timber, between 1900 and 1950 the 'real' costs of materials production have for some years been declining and this decline has helped our standard of living to rise. Nordhaus studied the relation of prices of ten important minerals to the cost of labour from 1900 to 1970 and found that 'there has been a continuous decline in resource prices for the entire century' (quoted in Barnett, 1980).

One reason that mineral prices were so low in the boom years of N. American industrialisation is that the true costs of producing and processing the minerals were nowhere reflected.

The consequences of ignoring the non-quantifiable 'quality of life' costs have since become obvious, but it seems that the true energy and water costs were not taken into account either. For example, to produce a tonne of copper requires the energy equivalent of some 18 barrels of oil; the energy cost component of aluminium processing is about two and a half times as high. None of these costs have been reflected in metal prices. As yet, we also have not taken into account the economic cost of foul air, ruined streams and polluted soil.

Between the years 1963 and 1968, there was a steady expansion in the world mining production industry, an expansion which averaged 4.8% per year for this period. Mining production expanded by more than 10% in Asia, by slightly under 10% in the 'developing market economies' such as Central and S. America, Africa and S. E. Asia, and by only 4% in Europe and 2.6% in the USA. This expansion continued until 1971 when the world mining industry went into an overall recessionary phase, see Figure 6.5.

The most significant contributing factor to this recession was the decline experienced in the economies of the 'developing countries'. The trend was slightly reversed in the 'developed market economies'. For example, in N. America a very sharp recovery was recorded, but this was offset by the continued

Figure 6.5. Index of world mining production. (*a*) World. (*b*) Developed market economies (N.America, EEC, S.E. Asia, Australasia). (*c*) Developing market economies (S. and C. America, Middle East, Asia). Source: *AMIA Rev.*

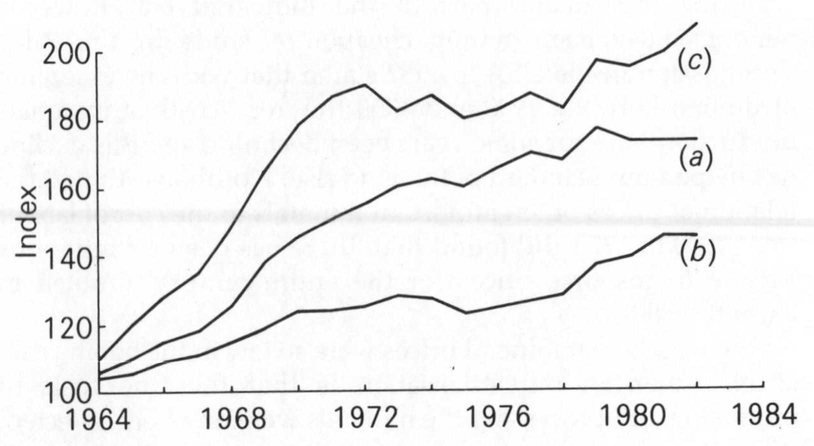

recession in Europe, and in particular the European Economic Community (EEC).

Various factors combined to contribute to this lessening of the overall growth of the world mining industry. These factors included many new restraints imposed by various environmental legislationary measures, new uncertainties imposed by currency realignments and inflationary wage pressures, especially within the 'developed market economies'. In addition, the general industrial unrest in the 'advanced countries' resulted in higher metal prices, a depletion of consumer stocks, and increased substitution.

During the period 1968–72, a period of change, the growth in mining production was most significant in Asia, which, excluding Israel and Japan, recorded a 41.7% increase. Growth was least in the 'developed countries', particularly within the EEC, which is heavily dependent on the import of raw materials. Amongst the 'developing market economies', growth was least in Central and S. America.

The indexes still reflect the continued recession in the Western economies since 1972. In particular there has been little, if any, growth in the mining index for Europe or N. America. Expansion of production continued in the petroleum producing countries of Asia and the Middle East following the production reductions of 1975. The Australian and New Zealand mining industry showed an overall revival in 1977 resulting mainly from increases in production of a large number of minerals rather than from large increases in one or two commodities.

The world production of some individual commodities is shown in Figure 6.6. Variations in the growth rates of production from commodity to commodity can be expected, but the extent of the individual variations within each year, and in the longer term, reflect the inter-relationship of the economic and technical factors on the industry. World production of all major commodities increased quite rapidly between 1964 and 1969; for aluminium the expansion and growth of production continued to rise quite steeply until 1974.

During the period 1969–71, there was a recession which adversely affected the output of all major metals with the exception of aluminium and tin. The decline in output was most particularly marked for lead and steel during 1970–1. A general recovery is seen to have occurred in 1972 in which the production

CARDIFF
UWCC LIBRARY

indices for copper rose by 9.6%, for steel 8.1%, and for lead 7.3%. Consumption of primary aluminium in the non-communist countries is estimated to have increased by almost 12% in 1972. The general revival in the world economy which became apparent in the latter half of 1972 resulted in increased consumption of steel, lead and copper. World production of copper rose by 8.3% in 1972 to almost 8 million tonnes, and the world output of steel grew 8% to 628 million tonnes. Increased industrial activity provoked shortages in many mineral products resulting in an upward pressure being placed on most mineral prices.

Since 1972, the world production of lead has stabilised at about 130 units. The production of copper has fluctuated from a low in 1975 – relative to 1972 – to peak in 1977 at just over 9 million tonnes. The world production of steel has continued on its downward trend since 1974, and by 1977 had decreased to 673 million tonnes. The demand for steel remained depressed throughout the period 1974–7, and a situation of oversupply dominated the world markets. Production of iron ore was therefore also lower in most countries due to depressed demand and surplus stocks.

World consumption and production of aluminium rose sharply until 1974 when there was a small decline in production above the 1973 level. Since then the output of primary

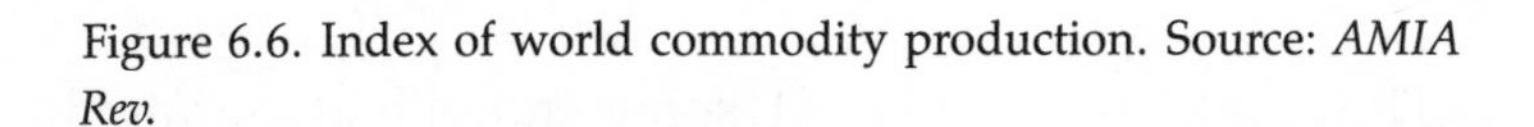

Figure 6.6. Index of world commodity production. Source: *AMIA Rev.*

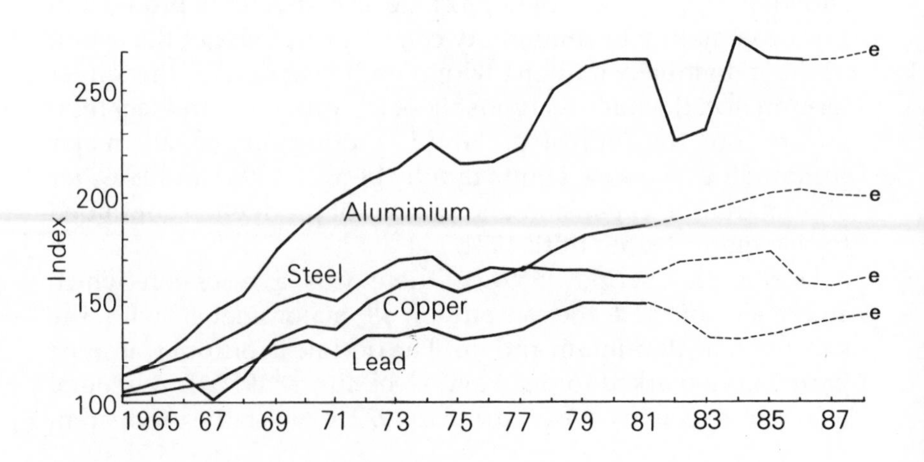

aluminium has once again increased. World production of primary aluminium increased by 8.4% in 1977 to 14.2 million tonnes: production in the non-communist world increased by over 10% to 11.4 million tonnes, compared with only a slight rise in total communist world output.

In the process of being consumed metals commonly flow from the production of primary and secondary (ex-scrap) metal through various stages of fabrication of intermediate alloys, to the production of final or end-use goods. The sum of the output of all final goods and services, equal to the sum of the value added (or income) from all the production sectors, is called the total final demand or the gross domestic product (GDP) (Etheridge, 1979, p.270). Etheridge states that: 'the demand for metals is ultimately derived from GDP which signals its appetite for metals through various intermediate metal processing operations back to the primary and secondary metal producers'.

Various attempts to measure the 'indirect' contribution that minerals make to an economy have been made by means of counting the jobs and income of associated input industries, and invoking a multiplier effect. The citing of multiplier effects is somewhat spurious in the sense that all industries that are integrated into a modern economy can claim a similarly enlarged contribution. Despite the occasional importance of minerals to a country's economy and balance of international payments, the possession and exploitation of mineral resources in themselves are in no way necessary conditions for a country achieving economic prosperity, which ultimately depends on the population rather than the resources.

Etheridge (1979) proposes that 'The changing level of metal consumption *M* in an economy is definable in terms of three key variables which dynamically interact through time with four conditioning variables. The three key variables are (i) the level and age structure of the population and workforce; (ii) the level of real GDP *per capita*; (iii) the real price of metal in relation to other goods and services, the metal's own past values, and to the prices of other metals and materials. The four conditioning variables are (i) the current level of technology; (ii) the timing of the path of the particular economy to industrialisation; (iii) governmental policies, especially with respect to taxation; (iv) local geographic conditions such as topography, climate and markets. Different economies therefore offer many contrasting experiences. For example, economies with similar real GDP *per capita* can exhibit many different patterns of metal consumption intensities, including differing structural and intensity characteristics. Now almost all the current methods in use for forecasting the future metal consumption trends are based in essence on exogenous (independent) assumptions regarding the future levels and patterns of both GDP (G) and population (P). Utilising historical data, mathematical correlative relations between the metal consumption (M) and G and P are developed most commonly by the calculation of various coefficients. The final stage is then to project consumption on the basis of these developed relationships, which sometimes are adjusted qualitatively.

'The types of models used to forecast generally fall into two categories. In the first sub-division or group, M is related in turn to the disaggregated components of G. The second group is characterised by the method of relating M directly to the aggregate economic variables such as G, P, or G/P. However, despite the complexity of some of the models none are necessarily reliable for a variety of reasons. For example, the behaviour of G, P and other factors is uncertain and not completely definable; and even when values are assigned to the exogenous variables such as G, or P, the models used are commonly not based on rigorous econometrically modelled relationships. In addition, many models resort to a simple extrapolation of observed historical trends, and on many occasions the forecasters make *ad hoc* adjustments. We feel that many of these problems result from a lack of appreciation of the true nature of the economic relationships. The interplay of the economic variables seems to be more stochastic than deterministic, and very little credence is given to the importance of conditioning. Conditioning relates to the probability of something occurring providing that a certain set of conditions are met. For instance, 'there is an 0.8% probability that the price of metal Y will be $3 per kilo provided that a GDP growth of 5% is maintained for the next year', is a simple conditioned statement'.

In this method the focus is placed upon the concept of metal consumption intensity M/G. The future expected metal consumption is related to real GDP, particularly in relation to the industrialised path of an economy as reflected in a rising *per capita* output, *G/P.* This inter-relationship is conditioned in turn by the impact of changing technology and relative metal prices.

In the early and middle phases of the path to industrialisation the economies see a rapid rise in metal consumption intensities. This rise is due to a rapid growth in the demand for the basic infrastructure goods such as electric power, water, railways, roads, cars, and general 'white' goods, and for metal intensive investment goods such as 'plant' and machinery. At this stage there is a high income elasticity of demand for the durable goods; that is, the demand growth tends to exceed the growth of the GDP. As the path of industrialisation proceeds, this income elasticity of demand for metals tends to fall; the growth of metal consumption falls relative to the GDP.

At even higher levels of GDP *per capita*, there emerges a high income elasticity of demand for, and a structural shift toward, low or non-metal-intensive goods and services. The recently industrialised economies like Japan, S. Korea, or Brazil, using the latest technologies available, have achieved much less metal intensive paths of industrialisation than did the older industrial economies such as Europe or N. America. Figure 6.7 shows the comparative timing of industrialisation for three countries; one 'old' economy and two 'new' economies, all of which represent the current markets for Australian mineral output. The data used are taken from Etheridge (1979). The GDP data used for all countries were adjusted relative to the USA to reflect GDP measured not in the common currency at prevailing exchange rates, but rather in terms of so-called comparative purchasing power parities. As plotted in Figure 6.7 the newer economies as represented by Japan, have achieved similar stages of development and industrialisation as measured by GDP far quicker than the older economies of the USA and Germany.

Figure 6.8 shows the changes in metal consumption and intensity for the period 1970–83. Consumption, with the exception of tin, has increased throughout this time period. Metals intensity, which is defined as the consumption per unit of GDP, deflates the reported consumption by the global growth in GDP of 51.5% during the same period. This tends to add weight to Etheridge's argument that the newer, mostly Asian, industrialised nations are following a much less 'mineral intensive' path to industrialisation.

Even so, based on current mineral usage rates, many tables of dynamic resource life have been published. One such table is shown in Table 6.1, which suggests that for some metals exhaustion of reserves should occur in the not too distant future. However, many commentators, including Beckerman (1977),

Figure 6.7. The path to industrialisation. (*a*) USA; (*b*) Germany; (*c*) Japan. Source: Etheridge (1979).

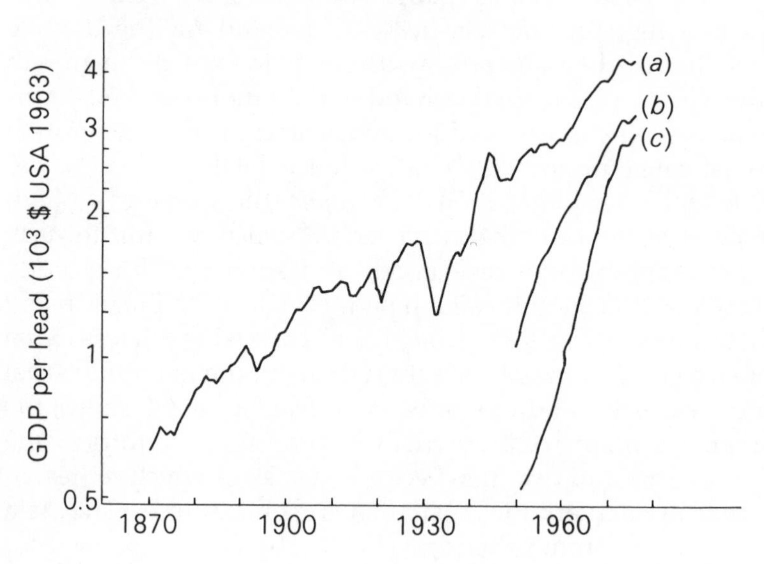

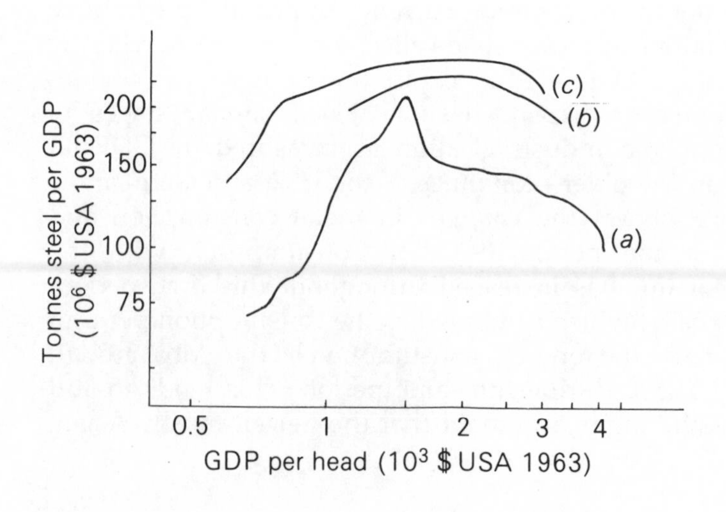

believe that not only is there no imminent scarcity of mineral resources in Australia but that, moreover, there is no likelihood of there ever being such a problem. The contention is that technological progress and economic adjustment prohibit such a state. They believe that technological advance acts in the same fashion for non-renewable resources as does the natural reproduction rate for renewable resources. Beckerman (1977, p.117) states: 'Thus as a matter of straight historical fact, however fast demand has expanded in the past, and for however long, new mineral reserves have been found, or some other painless adjustment process has taken place'.

To substantiate this point that fear of resource exhaustion is not new, Beckerman quotes the historical reserve situation for coal, lead and tin, and Barnett (1979, p.7) quotes the case for oil. The basic tenet of their argument is that at various times past finite reserve situations have been quoted, yet economic adjustments such as price increases and technological advance have always sustained a reserve no matter what the demand. Barnett further contends that technological progress has permitted, and will continue to permit, the development of lower-grade reserves, and cites the case that it is now often cheaper to mine low-grade copper deposits than to operate high-grade, smaller tonnage mines.

Figure 6.8. World metals consumption and intensity 1970–83. Source: *The Age* (27;12;84).

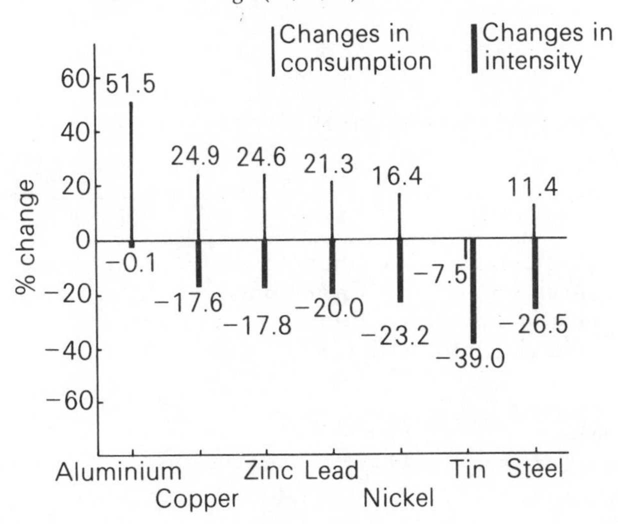

But as Govett & Govett (1974, preface) state:

> 'The tendency to view the problem of resource adequacy from the politicians' and economists' point of view – while understandable – results in studies which list a number of options for society based on technological and geological assumptions which are not necessarily either correct or meaningful or which have been taken out of the original scientific context . . . This tendency to assume that science and technology will provide the answers whatever scenario is adopted by national and international planners is dangerous'.

Table 6.1 *Dynamic resource life (years)*
Source: Ray (1984)

30–100	100–200	200–400	400–800
Gold	Copper	Aluminium	Manganese
Tungsten	Nickel	Chromium	Titanium
Lead	Platinoids	Iron	Potash
Zinc	Molybdenum	Cobalt	
Mercury			
Silver			
Tin			

7

Mineral resources in the global context

The changing flows in world metals trade are shown in approximate terms in Table 7.1. As the figure shows the once very considerable sectoral concentration within the industrialised world has changed. The non-industrialised, or essentially 'third world' countries, once only the suppliers of raw materials to the major consumers, are becoming consumers themselves. Now although some countries are in the fortunate position of balanced imports and exports, there are those countries that must import all of their raw materials and those that export all of the raw materials production. Australia could be an example of the first type, Japan an example for the second case and Chile or Zambia an example of the last case.

In the resource importing countries which are bereft of natural resources, such as Italy or Japan, the high levels of industrial output are achievable only at a high cost. Problems of ever-increasing costs, inflation, very high levels of pollution and environmental degradation coupled with periodic shortages of materials must all be countered so as to achieve continuity of output. By contrast the resource rich countries, especially those with petroleum based wealth, are placed in a very strong global manipulative position in that they can exert a degree of control over the process of 'growth and development'. Resource poor countries will continue to be bereft of bargaining power since rising costs of raw materials and manufactured goods must inevitably continue to feed real income from resource poor to resource rich nations. It is interesting to note that the global picture has not changed appreciably in over a decade.

The underlying thesis for this concept of mineral resource utilisation is that the global material civilisation depends, in the main, on the continued use and further discovery of mineral resources. The physical structure of the system is simply that materials are taken from the Earth's crust and posited as a pool of available supply. These supply phases are of course very dependent upon an energy base to sustain them. After purchase the materials remain 'in service' for varying lengths of time after which they are either reused (recycled) or dumped. Efficient recycling systems are desirable, if for no other reason, because they increase the dynamic resource life of the material. Naturally this then reduces the pressure of demand upon the primary reserve base. If the industrialised nations initiated high efficiency recycling systems an unfortunate consequence would

Table 7.1 *World metals demand 1900–80 (million tonnes)*
Source: SRI International (1985)

				Average annual growth rate (%)	
	1900	1920	1980	1900–50	1950–80
Steel	41.0	191.60	715.90	3.1	4.5
Copper	0.513	3.01	9.42	3.6	3.9
Lead	0.871	1.87	5.39	1.5	3.6
Zinc	0.475	2.08	6.17	3.0	3.7
Aluminium	0.007	1.58	15.30	11.4	7.9
Molybdenum	—	0.02	0.80	—	5.7
Nickel	—	0.16	0.71	—	5.2

	Regional metal consumption (% world total)					
	Western industrial nations		Developing nations		Eastern bloc	
	1950	1980	1950	1980	1950	1980
Steel	76	59	5	16	19	35
Copper	84	67	4	10	13	24
Lead	84	62	6	11	10	27
Zinc	85	58	4	14	12	28
Aluminium	82	69	2	10	16	22
Molybdenum	85–90	73	NA	5	NA	22
Nickel	81	68	NA	6	19	26

be the disabling of many third world economies who are dependent upon mineral resource exports as hard currency sources.

The materials system itself of course has to complement the social control which is manifest upon the mineral resource trade flow. Naturally consumption determines the rate at which materials flow through the system. The magnitude of the demand for any particular mineral is dependent upon the interrelation of three factors: the size and growth rate of the population, the level of perceived need for the particular mineral and lastly the price of that particular mineral as against the cost of alternatives and substitutes.

As material civilisation continues it is possible that the requirement for mineral resources will not only continue but escalate. It is therefore pertinent to take stock of the available minerals on a global basis. Table 7.2 outlines the pattern of estimates of reserves available at differing time periods for several key metals. These estimates are of course restricted to known occurrences and reflect the current level of technology and expertise in extraction and development. It seems reasonable to assume that whilst large areas of S. America, Antarctica, Asia and Africa remain essentially unexplored global reserves of most of the metals will continue to increase. As long

Table 7.2 *Estimates of global reserves for 'key' metals*
Adapted from McLennan (1978)

Metal	1950 (*a*)	1960 (*b*)	1976 (*c*)	1980 (*d*)
Copper	196	245	459	543
Lead	40	44	145	126
Zinc	72	76	159	150
Tin	6	NA	10	10
Nickel	14	14	55	54
Iron	56	84	93	258
Aluminium	500	900	4300	15400

Values in million tonnes: iron values in billion tonnes
(*a*) Paley report
(*b*) Ford FDN: resources in America's future
(*c*) USA bureau of mines
(*d*) Bureau of mineral resources, Australia

as demand continues to encourage exploration this situation will continue, and will probably exist at least until the 21st century.

Mineral resources are not, however, evenly distributed throughout the world and not a single nation is able to claim total self-sufficiency although some, such as Australia, come fairly close. It is generally assumed that the resource rich countries have an obligation to develop their reserves for the greater benefit of mankind. The argument commonly takes the form that the host country benefits from an increase in export earnings as well as having a stimulus given to its own local industry. The mining industry is argued to be a promoter of decentralisation and the provider of provincial infrastructure within the host country. A free market does not, however, exist.

For many minerals there are producer cartels which aim to provide price stability by controlling mineral production. This is attempted by developing a system of maximum and minimum prices backed by a system of production/export quotas and buffer controls. The success of these producer cartels is decidedly arguable; many have been abject failures, for example the tin cartel.

Consumers have also established their own cartels aimed at stabilising prices. Joint negotiation and purchase has enabled the controlled development and usage of resources commonly to a significant advantage. The Japanese steel industry is a good example of this.

Oft times resource diplomacy is utilised to achieve an international or national political gambit. Government utilises its mandate to control raw materials, or processed goods, so as to counter particular economic conditions regarded as deleterious to its own well-being. Similar actions have been taken from time to time by minority group action. Such actions have had varying degrees of success. The overriding point, however, is that a free market simply does not exist in the global mineral industry.

7.1 Global resources – the Australian perspective

Australia is essentially self-sufficient in most minerals and hence it is a major exporter in the global context. However, Australia does not have a monopoly in any one mineral. In particular, Australia is a supplier of coal, bauxite and alumina/

aluminium, iron ore, copper, lead and zinc to the global market. The following is a brief review of these commodities and is taken from the *Mining Review,* May 1984, pp. 17–20.

> 'Of total world black coal production (about 2.8 billion tonnes) only about 230 million tonnes enters international trade. The USA, the People's Republic of China and the USSR are the major producing nations with production levels ranging from 500–900 million tonnes. The combined reserves of these countries comprise about 70% of known world reserves. The USA is both the main producer and the main exporter.
>
> 'Transport costs are a major factor affecting coal's competitiveness. There are also cost penalties which do not apply to other energy forms. These are mainly related to extra investment in handling and pollution control equipment.
>
> 'Coal has been partly insulated from the effects of the world-wide recession because of the relative price advantage it enjoys over its major competitive energy source – oil.
>
> 'The International Energy Agency (IEA) and the World Coal Study (WOCOL) envisage an important role for coal-supplying countries such as the USA and Australia in the enlarged coal trade of the future.
>
> 'According to the IEA coal use in the 24 OECD nations will increase from 1.16 billion tonnes in 1980 to 1.25 billion tonnes in 1985, 1.5 billion tonnes in 1990 and 2.3 billion tonnes in 1995.
>
> 'The three biggest producers and consumers will remain the USA, China and the Soviet Union, but the biggest production expansions in the next 10 to 15 years are expected to occur in smaller countries like Australia.
>
> 'By 1985 the IEA predicts that the significant nett exporters of coal will be the USA, Australia, Canada, South Africa, Poland and the USSR.
>
> 'The world aluminium industry is characterised by a high degree of concentration and vertical integration. The six largest companies control 70% of alumina capacity and 60% of aluminium capacity. These are Alcoa, Kaiser and Reynolds (all of the USA), Alcan of Canada, Alusuisse of Switzerland and Pechiney Ugine

Kuhlmann of France. All these firms are also involved in the Australian industry. Vertical integration usually encompasses smelting and semi-fabricating/fabricating with strong involvement by the major metal producers in bauxite mining and alumina refining.

'The USA is the largest world importer of bauxite and is a leading alumina producer and importer. Other major importers of bauxite are Japan, FR Germany, USSR, Italy, Canada and France. In the case of alumina, major importers are Canada, Norway, Japan and the UK. Japan, the USA and FR Germany are the largest importers of aluminium.

'The sale of bauxite and alumina is covered by contracts ranging in duration from a few years to several decades. Long-term, large tonnage contracts are often between related parties and are designd to underpin investment in mining and refining facilities and to secure ongoing guarantee of availability.

'The world iron ore market is notable for the fact that, with the exception of the USSR production, a large proportion of iron ore is traded internationally.

'The steel industries in Europe depend on imports for most of their iron ore needs. All of Japan's iron ore is imported. Of the major steel producers, the USSR and China depend mainly on local iron ore production, while the USA imports a quarter of its requirements, mainly from Canada and South America.

'In recent years there has been a move to greater dependence on a few relatively large producers of iron ore. Ten years ago Japan relied on 24 countries for imports of iron ore; today only 13 countries supply Japan with iron ore and almost 94% is supplied by five countries. Seven countries – Australia, Brazil, Canada, the USSR, Sweden, India and Liberia – account for 80 per cent of the world export trade.

'The steel industries in Europe and Japan have developed a fairly stable pattern of sources for their supplies of iron ore. European steel makers depend mainly on suppliers in America, Africa and Sweden. Japan depends primarily on three countries for its supplies – Australia, Brazil and India. Increasing freight rates are causing steel makers in Europe and Japan to give greater emphasis to supplies of iron ore located nearer to their plants.

'The trade in iron ore is characterised by long-term contracts with regular, mostly annual price reviews. The move towards reliance on a smaller number of large suppliers is expected to continue. This is because iron ore mining has become highly mechanised and dependent on large throughput to achieve economies of scale.

'Increased bunkering costs arising from rises in international oil prices in recent years have enhanced Australia's advantage as a supplier to Japan and the developing markets in Asia and the Middle East.

'The improved results were largely due to the higher contract prices negotiated with Japanese mills in 1982 and devaluation. In late 1982 and early 1983 the focal point was the problems the Pilbara iron ore miners and Australian eastern seaboard collieries were having maintaining markets and contracts, particularly with Japan.

'Resources of lead and zinc are spread widely throughout the world and, in most mines, the minerals occur conjointly, but in varying relationships.

'Major producers of lead concentrates are the USA, Australia and Canada, which together account for approximately 50% of western world supplies, together with other significant production in Mexico, Peru and Western Europe. Large quantities are also produced in Centrally Planned Countries, with mine production in the USSR estimated at 577 000 tonnes in 1982 (slightly higher than US production).

'Refined lead production is widely spread with recycled scrap lead representing a significant proportion of lead consumption. Main uses of lead are in batteries, chemicals, cable sheathing, petrol additives, pipe and sheet. For zinc, main uses are for anti-corrosive treatment, diecasting metals and for alloying brasses and the like.

'Technological change and environmental factors are influential in the demand for both metals. Lead is being phased out of petrol and has been replaced by plastics in some applications. On the other hand, increased use of lead is likely for energy storage in batteries. While the use of zinc has been greatly reduced in motor vehicles, the advent of thin wall diecasting could reverse this change.

'Canada leaves western world zinc production (17.7% with Peru, Australia (each accounting for around 8% of world production), the USA, Mexico and Japan also contributing significantly. Leading metal producers are the major industrialised countries and Australia. The USSR is believed to be self-sufficient with production and consumption estimated to be a little more than one million tonnes a year (16.3%).

'The major consumers of lead and zinc metal are the leading industrialised countries.

'Lead and zinc markets are not necessarily both strong or weak at the same time. Since these minerals are produced conjointly, producers must determine optimum output level and the stronger metal must often subsidise the other.

'Australia is a foundation member of the International Lead and Zinc Study Group which comprises all major producing and consuming countries. The Study Group meets annually to review the world market situation and undertake special studies from time to time. In its 1983 report the Study Group said there were now another eight countries producing over 100 000 tonnes a year, compared with just two in 1960.

'The USA overtook both Australia and the USSR as the biggest producer with the opening of major new mines in the Missouri lead belt in 1969, before losing its lead to the USSR in the mid-1970s'.

7.2 The Pacific scenario

The Pacific Basin contains a substantial proportion of the global mineral reserve. The Pacific Basin is a term used to denote the countries with Pacific Ocean shorelines. The Western Pacific Region is that area to the west of the International Date Line, and the Eastern Region to the east. An indication of this abundant reserve is shown in Table 7.3. Whether or not these resources will be developed in the near future will depend upon the close co-operation of the various governments in the region. There is, however, sufficient resource to posit the formation of a Pacific Economic Community. Before such could eventuate it would be necessary to identify the policy opportunities and the negative aspects of such so as to facilitate the full development of the mineral resources in the region.

The Pacific region is one of the most prospective regions in the world with only a small part of the landmass having been subject to modern exploratory surveys. The ocean itself is a repository of mineral wealth which at this current time is only being partially developed. Historically mining on a large scale is a relatively recent phenomenon in the region and the landmass has not been the subject of years of intense exploration effort as, for example, in N. America or Europe.

A mineral resource is a wasting asset and this provides the stimulus to encourage the search for new replacement mineral deposits. The cost of finding minerals is very high since most of the easy-to-find deposits have already been discovered and developed. Most effective mineral exploration budgets are now costed in terms of millions of dollars with a very high risk factor. Collier (1983) quotes research that shows that the cost of finding an economic mineral deposit in Australia averaged about $40 million dollars for the period 1964–78. The cost of development has also risen considerably in the last two decades. The cost of

Table 7.3 *The mineral resources of the Pacific region*
Source: Rutland (1987)

Metal	% global reserve
Coal	25
Tin	59
Nickel	30
Zinc	21
Lead	14
Iron ore	27
Bauxite	28
Uranium	16
Copper	40
Silver	44
Tungsten	59

development is now reckoned in terms of billions of dollars to whit Bougainville and OK. Tedi. Much of this cost is a factor of remoteness of location, inclement climate and environment and the high cost of infrastructure provision. Risks of development are also increased by the occasional desire for local expropriation to enhance domestic foreign currency earnings. All of these are essentially the result of local policy positions.

The Western Pacific region alone covers an area of some 21.6×10^6 km^2 with a population of about 1.5 billion – about 15% of the global land area and 30% of population. During the 1970s economic growth in this region outpaced global growth. During 1970–9 the world economic growth averaged 4% *per annum*; this growth rate was exceeded by 13 of the region's economies. The total GDP for this region increased from 12% of global GNP in 1970 to 16% of global GNP in 1978. Despite the decline in world economic growth rate to about 1% in 1980 with a concomitant drop in world trade volumes (6% in 1979 to 1% in 1980) economic growth in most of the region has remained strong.

The Western Pacific region alone accounts for about 16% of global exports, of which some 6% is intraregional trade. 38% of exports are directed to countries within the region and 39% of imports are sourced within the region. Japan holds the dominant trade position. The regional trade pattern is a reflection of the relative resource and technological capacity of the individual countries.

China's energy reserves are the largest in the region with significant reserves of coal, oil and gas. Uranium reserves in China are unknown. Indonesia, Brunei and Malaya are very significant oil, gas and LNG producers and exporters, whereas Thailand's natural gas reserves are mainly utilised in satisfying domestic demand. New Zealand has natural gas resources which it is currently utilising as CNG and LPG for automotive purposes as well as for feedstock for methanol and synthetic petrol production. Australia's reserves of the more important metals are adequate to sustain forecast rates of production well into the 21st century.

The socio-political and economic factors affecting demand and supply such as the political and military conflicts between countries, the formation of trading blocs and the natural cyclic character of economic activity will continue to influence the market structure. Whether the region does fulfil the potential which the resources would infer remains to be seen: political stability and co-operation remain the major deciding parameters.

8

Political, social and economic factors in mineral resource systems

Our global societies are more or less mineral societies dependent in large manner upon the consumption of large quantities of minerals and mineral products. Mineral consumption in the past 50 years has been greater than for the whole of recorded history. The unfortunate aspect of this perhaps is the fact that most of this has been concentrated amongst a handful of nations who now represent the industrial and technological elite. The consumption *per capita* in the USA is equivalent to about 8 t of fuel (in terms of oil equivalent), some 4.5 t of sand, gravel, salt and sulphur, and about ¾ t of metal – predominantly steel with lesser amounts of aluminium, copper, tin, lead and zinc.

The desire for nations to aspire to these levels of consumption has given rise to the large-scale global mineral demand. The scale of resource demand has also given rise to fears at various times past of possible exhaustion in the not so very distant future. In addition to fears of exhaustion the past two decades have seen the emergence of expression of two further concerns: environmental degradation and pollution.

The disruption of large tracts of land as the result of open-cut or strip mining, predominantly for coal, has caused great concern to many who see this as an unnecessary and wasteful destruction of the environment. Even the introduction of large-scale rehabilitation programmes have not quelled the alarm or satisfied the environmental lobby. Examples are the Darling Range bauxite operations in Western Australia and the open-cut coal mining areas of central Queensland.

The industrial practice of dumping wastes in convenient rivers, streams, lakes and seas, which has gone on for many years, has also recently been the subject of expressions of alarm and grave concern. Pollution, with all of its disastrous consequences, mutated farm animals, human sicknesses, and toxic water, have not endeared industry to suffering populations. The fear of radiation due to reactor failure has probably been the greatest and most universal expression of distaste and concern. These factors give rise to conflicts over competing land use options for which the mining industry is commonly regarded as an aggressive despoiler. An example in Australia would be the coal developments in the Hunter Valley of New South Wales and the attendant proposals to build aluminium smelters. Much debate was therefore centred on the proposals to mine and smelt regarding the disruption to the cattle and viticulture industries by fluoride emissions and various sundry attendant pollutants.

More commonly, however, mining operations have been sited in the more remote regions of the world which have not been subject to competing land use. This has also meant that throughout the ages the general population has had little knowledge of the mining and mineral processing industries. It is only in the past 50 years as mining has taken on an increasingly corporate nature and head offices have been established in the major cities that public interest has arisen. This has meant that the industry has been subject to an increasing amount of legislative and public pressure to ensure that it meets the current social, political and economic dictates of government policy. The mining industry, especially within the framework of Western industrialised nations, is being increasingly affected by decisions taken in what are termed the public interest. The industry therefore must be considered today in its total social and political framework as well as in its more traditional economic setting.

There is a substantial literature in respect of attempts to place the mineral industry into some form or perspective with respect to long-term supply and ready availability of mineral commodities. This was touched on in previous chapters. These descriptions will never be more than partially successful since there are numerous uncertainties that surround the definition of available reserves, the variable boundary defining reserve as distinct from resource, and the consumption patterns of the future. Periodically, predictions of doom have been made and at times and in various locales shortages have been experienced, global shortages in terms of impending exhaustion have not, as yet, come to fruition.

The policy settings and considerations of the industry's decisions, whether they relate to the company, the nation, the cartel or the political–economic bloc, have undergone considerable metamorphosis over the past few decades. One of the emergent factors has been the rise and increase in economic power of the multi-national mineral enterprise wherein one company grouping has control over the supply, the processing and the sale of a commodity or commodities. Aluminium is of course the type example.

Another problematic consideration in terms of forecasting the future relates to the raw materials requirements of third world countries as they pursue the path to industrialisation. As their standard of living moves toward greater levels of mineral use the global supply situation could become more acute. Considerations such as rising prices, âbilities of poorer nations to pay, and the likely affect the third world demand will have on the raw material position of the technologically advanced economies all come into the analytical framework. The actual analysis or forecast projection will depend upon the assumptions made about the technology to be developed, the economic regime and the political stability of the nation or region, and the time scale and geographic interplay. The projection or scenario will therefore vary according to the number of minerals considered, the number of users considered and the number of suppliers considered. Despite the vast number of variables there are some concepts that are fundamental to our understanding of the mineral resource system.

Firstly it is generally accepted in the industry that mineral resource availability is mainly a matter of cost rather than of the possible finiteness of the reserve and/or resource. Past civilisations have not 'run out' of a particular mineral and neither does it appear that any civilisation has been limited in its development because of perceived shortages of any one mineral commodity. Perceived shortages in the past have been due to very rapid, timewise, surges in demand, rather than due to any basic lack of reserve. Recent shortages have been essentially 'local' and resulted from changes in domestic laws: health, safety, environmental and economic. The actual reserves have, as was shown in the preceding chapter, increased as exploration has covered new virgin areas and also increased in sensitivity.

Mineral availability will not increase forever, a finite limit does exist. However, some counter to ever-increasing utility does exist. Apart from the researches into better extraction and exploration technologies that could provide greater reserves by definition of new ores and/or lower grades, demand will be constrained by the development of substitutes or alternate methodologies. New ideas may eliminate some mineral usages completely. Recycling may be increased. Mineral availability is not really a question of imminent shortage. The length of the time horizon into the future is what determines the actual scenario and interplay of the above parameters.

In a market or mixed economy for a deposit to be developed it must satisfy the grade, size and location conditions such that it can be mined at a profit. The cash flow and cash return – the economic decisions – are the defining criteria. In a fully planned economy it is the political objective that plays a more decisive part in the exploitation of a mineral resource.

The geographic spread of mineral resources throughout the world is not necessarily one that any particular corporate group or nation would choose. It is a spread that covers a great diversity of cultures and political groupings and one that ensures that the maintenance of industrial societies necessitates a costly transportation system. This allows the usage of minerals as weapons of economic trade, warfare and sanctions. Historically this has not always been the situation. For example, it is only since the Russian revolution that the countries of the communist bloc have identified self-sufficiency as a major motive in resource development. Similarly as a result of differing political persuasions being taken up, mineral trade can be significantly altered. As political change occurs so do mineral patterns of development and trade also change. The varying character of manganese trade since the 1940s is an example of this phenomenon.

Other examples which can be cited include the copper province in S. Central Africa, which now lies on the border of Zaire and Zambia. The industrial complex of the Ruhr based on the coal reserves and iron deposits of Alsace-Lorraine is another example of a mineral region which has been subject to varying political ramifications, to whit the past enmity between France and Germany.

Thus, although the location of mineral deposits are fixed this rigidity is a relative rather than an absolute characteristic. Shifts in political boundaries affect patterns of mineral supply and development. Changing technologies have meant that there are now far more mineral deposits than would perhaps have been anticipated, say, 50 years ago. In market economies the practice of discounting and evaluating mineral reserves of the future means that the true indication of a nation's mineral reserve/resource is not necessarily fully realised. In planned economies wherein the state carries the burden of high risk exploration and development there is commonly a tendency to prove up reserves as a political expedient rather than of necessity. Current indications of reserves and resources are often therefore the best guess at a 20 year inventory.

To be of use a mineral must be moved from the deposit into the general industrial flow of raw materials. For conditions of trade between nations of similar political, social and economic persuasions the problems are simply those of applicable technology and cost. In many ways, given the global political complexity, the movement of minerals from points of origin to the market place is a fairly smooth transition, even though the routes are more complex and costly than if there was a single global functioning political unit.

The uncertainties regarding the stability and continuity of the mineral resource system in the future reflect the uncertainties that are inherent in the geology, the economics, the politics, technology and environmental issues that pertain to each political unit.

Currently, the 'conventional wisdom' regarding the minerals system that we have described is undergoing rapid change. New technologies such as fibre optics and high temperature ceramics are causing major displacements of traditional metals such as copper, lead and zinc. Figure 8.1 very graphically portrays the increasing role of metal substitutes.

Figure 8.1. Historical perspectives – the new materials. USA production of aluminium and synthetic polymers. Adapted from SRI International March 1985.
Note: The usage of plastics has doubled approximately every four years since 1940 (~18% growth rate). About 3 million tonnes of coal equivalent are required to produce 1 million tonnes of plastic.

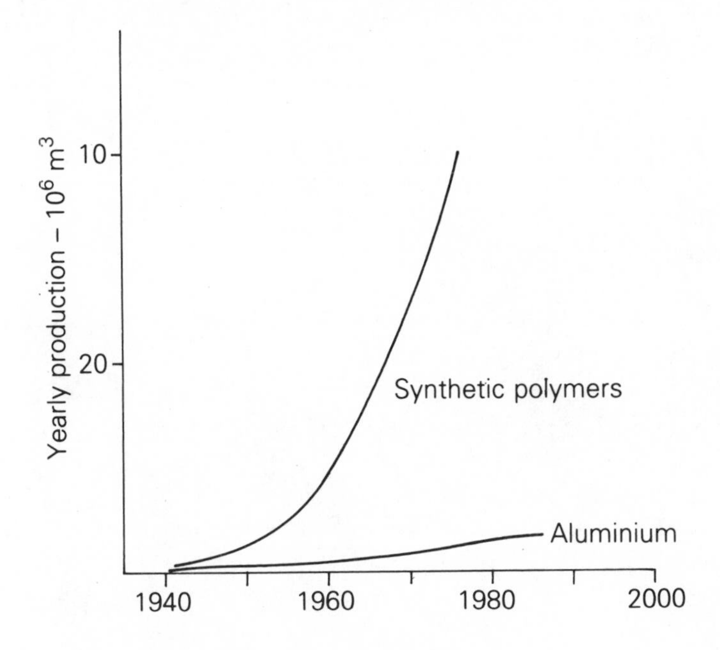

Part 3 Resource evaluation

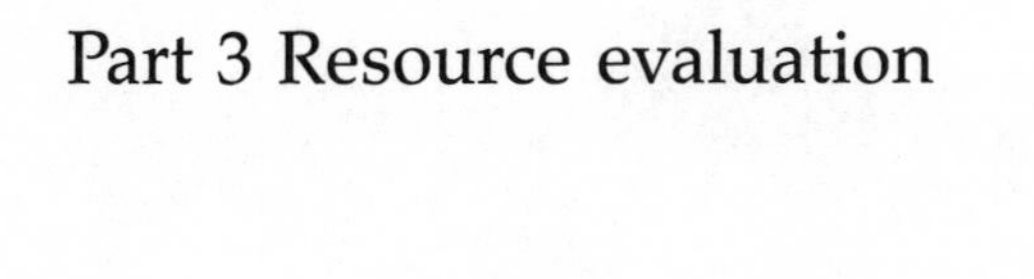

9

Exploration technologies

The search for and the finding of new mineral resources is an ancient art which has been effective for well over 2000 years. The traditional methodology has been prospecting, whereby individuals have used a random search philosophy, looking for 'kindly country'. The prospector was responsible for the great gold rushes of the USA, S. Africa and Australia during the 19th century. The effectiveness and reliability of the prospector has declined with the diminution of those deposits commencing at grass roots.

It was not until the late 1860s that the professional mining engineer entered the field of operations, and the professional mining geologist in the early 20th century; the latter was the resultant of legal requirement. This effectively remained the *status quo* until just after the Second World War when the mineral exploration function underwent a complete review.

The 'uranium boom' in the USA encouraged the organisation and development of literally hundreds of small companies who went in search of uranium deposits and quick profits. Concomitant with this was a growing concern felt by both government and industry that known reserves of various 'essential' metals were not sufficient to meet the projected demands. The picture painted was one of almost universal shortage of nearly all metals. Another significant factor which determined the review of mineral exploration methodology was the very real success that petroleum companies were having in the worldwide 'saturation' style of exploration.

'Saturation prospecting' for base metals came into vogue in the 1950s. Co-ordinated search techniques were developed for the recognition of the characteristics of various types of ore deposits. This approach to exploration yielded many major discoveries and was quickly adopted by many companies. By the beginning of the 1970s this technique had reversed the mineral reserve picture; universal shortage had moved to universal surplus.

During the 1970s the blanket search method of exploration declined due to escalating costs and a decline in discovery rates. This period saw the development of yet another technique, the genetic model approach. Models were developed defining particular environments for specific ore types and the search was then concentrated on defining and delineating the relevant geology.

Mineral exploration moved into the area of the specialist team effort. In Australia five examples of this type of exploration are readily evident: the Bougainville porphyry copper deposit; the Duchess phosphate deposit; the Argyll diamond deposits in the Kimberleys of Western Australia; the Roxby Downs uranium copper discoveries; and the porphyry copper–gold deposits near Parkes, central New South Wales. For each case the geological environment and characteristics were defined prior to the field exploration proper, Haynes (1979).

The time–exploration changes reflect a changing philosophy and approach to mineral exploration. The recent advances in ore petrology and theories of ore genesis and deposition have led to concepts of target generation within regional and continental contexts. In this sense, target generation is governed by such concepts as:

- metallogenic provinces;
- tectonic domains;
- past production records;
- value judgements as to risks.

Concomitant with this swing to a regional approach and large-scale surveys the involvements of large amounts of capital has promoted interest in methods of optimisation of the allocation of the resources of the exploration companies. Currently, most exploration in Australia is performed by large corporations, mostly multi-national, who have committed vast amounts of money, men and equipment to the search for fuels and mineral wealth.

The patterns of successful metallic mineral exploration in Australia for the past three decades (1951–80) have been analysed by several workers; see Haynes (1979), White (1979) and MacKenzie (1981). Several distinct trends are readily apparent. Increasing numbers of 'blind' or 'poorly expressed' ore bodies have been found, and this trend can be expected to continue. The number of discoveries utilising what can be termed traditional geological techniques have proportionally declined, but this exploration philosophy will continue at a significant level for several years into the future. Successful discoveries have also been aided by significant improvement of the quality and number of geological exploration models and the available basic geoscience data in more recent years.

The changing methodologies in mineral exploration in Australia are shown in Figure 9.1. In respect of the discoveries of deposits with good outcrop and surface manifestation there is a decline from about one-half the total in 1951–60 to less than one-quarter in 1971–80. This decline will probably continue. During the 1970s the discoveries of deposits with poor surface manifestations exceeded the 'good' outcrop discoveries for the

Figure 9.1. Changing exploration methodologies (Australia). Source: MacKenzie (1981).

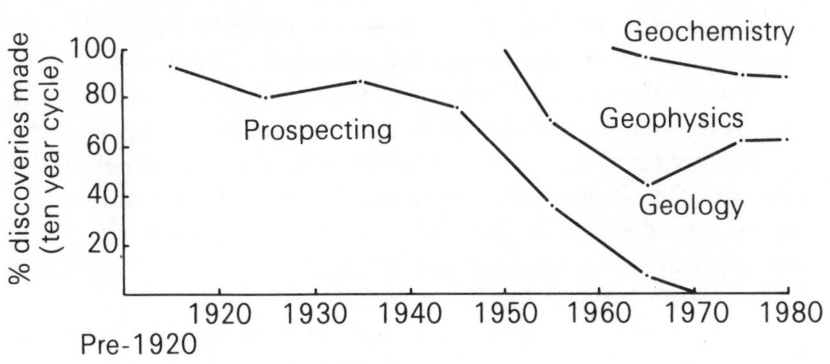

Surface expression of base metal discoveries 1951–80

	Outcrop good	Outcrop poor (but observably distinct)	Outcrop obscured (heavily leached, partly concealed, subtle)	Blind	Total
1951–60	9	4	2	2	17
1961–70	22	16	8	10	56
1971–80	10	16	7	12	45
Total	41	36	17	24	118

Targetting method of base metal discoveries 1951–80

	Prospecting (geological)	Conceptual geological models	Geochemistry	Geophysics	Total
1951–60	10	3	—	4	17
1961–70	30	7	5	14	56
1971–80	22	6	5	12	45
Total	62	16	10	30	118

first time as also did 'blind' discoveries. Discoveries of 'blind' deposits made a dramatic 'leap' in the 1960s due mainly to the better application of geophysics and to a lesser extent geochemistry.

The development of various conceptual geological models has also made a distinctive input into the mineral exploration format, especially during the 1960s and 70s. Some of the better known models which have been applied in Australia with sucess during these two decades include:

- bauxite on Tertiary to Recent lateritic peneplains;
- massive hematite in Proterozoic banded iron formations;
- nickel sulphides adjacent to ultramafic flows;
- uranium in Palaeozoic acid volcanics;
- shelf carbonate-hosted lead zinc deposits;
- poly-metallic massive sulphides in altered acid volanics.

MacKenzie (1981) gives a more complete listing. The trend toward more and better models will undoubtedly continue. As Haynes (1979) pointed out, Australia has great tracts of 'concealed' and therefore unexplored areas, and the application of models to concealed ground will require the most innovative, unorthodox and persistent geological approach. The emphasis upon geological technology and the reliance upon conceptual modelling can be expected to increase.

Haynes goes further in his analysis when he suggests that 'Geological technology has contributed directly to the discovery of most of the mineral resources found in Australia over the past thirty years' (Haynes, 1979, p.75). His concept of geological technology consists of two parts: the first is a conceptual model – or set of models – describing the geological environment of the mineral deposit, or deposits, of interest. The second part is the application of the conceptual model as a target selecting device in the exploration programme.

A definition of geological technology therefore can be given in that it is simply the development and application of geological concepts. In the Australian context effective prospecting of large areas has required the usage of relatively high cost geochemical and geophysical techniques. Such programmes are much more costly than programmes employing similar techniques within targets selected by conceptual models. An outline of the model development is given in Figures 9.2 through 9.4.

There are of course, several difficulties associated with the development of conceptual models. A major problem is that the number of fully documented geological case studies of 'type' mineral deposits is not generally sufficient to allow development of completely definitive models. The generally poor standard of communication between earth scientists is also a complication, added to which there is little co-ordination of

Figure 9.2. Geological technology 1. Development of geological models. Adapted from Haynes (1979).

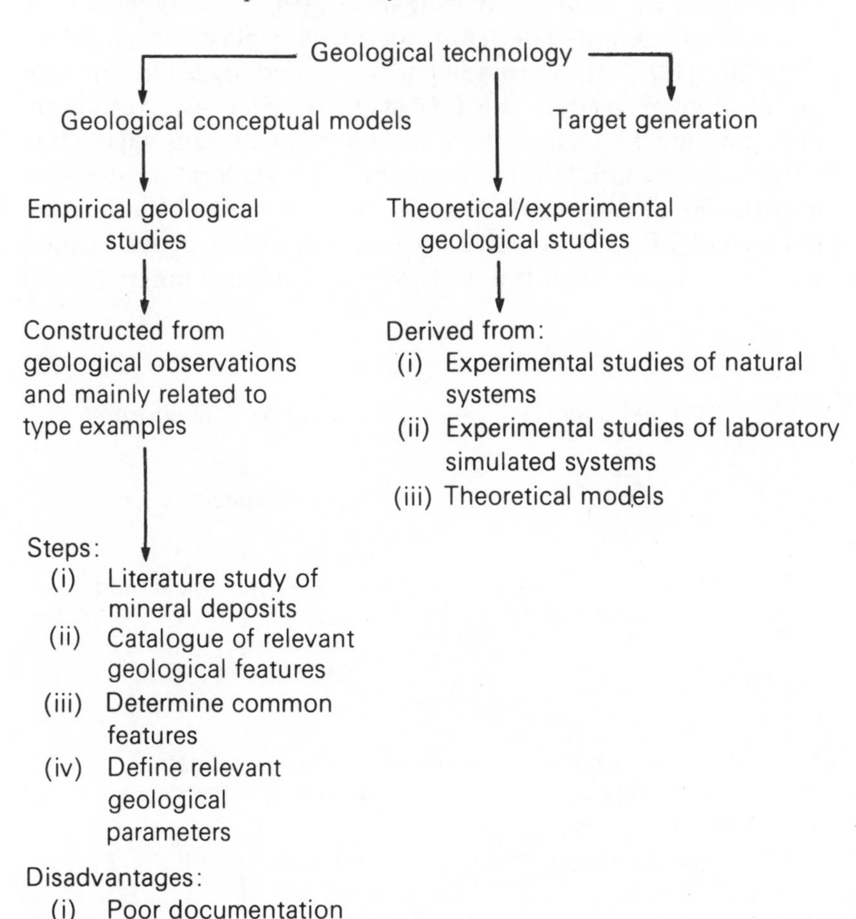
Geological technology
Geological conceptual models
Target generation
Empirical geological studies
Theoretical/experimental geological studies
Constructed from geological observations and mainly related to type examples
Derived from:
(i) Experimental studies of natural systems
(ii) Experimental studies of laboratory simulated systems
(iii) Theoretical models
Steps:
(i) Literature study of mineral deposits
(ii) Catalogue of relevant geological features
(iii) Determine common features
(iv) Define relevant geological parameters
Disadvantages:
(i) Poor documentation
(ii) Absence of standardisation
(iii) Variable data density

research between geoscientific groups in Australia. Finally, our knowledge is still incomplete; large gaps still exist in terms of theoretical and experimental data on mineral deposits.

As the difficulty of discovering new deposits increases, the concept of operations research in the minerals industry is becoming more important. Many workers, including Koch & Link (1971) have divided the exploration phase into several stages. Rendu (1976), recognises five stages: (i) exploration feasibility planning; (ii) regional exploration; (iii) follow-up exploration; (iv) detailed investigations; (v) ore-body valuation. A statistically oriented decision criterion is applied to each stage.

Rendu (1976) has provided a simplified decision tree for exploration problems. The first stage is when an exploration decision e must be made, the last stage is the mining stage when a mining decision M must be made. The exploration decision may be one of two possibilities, e_1 or e_2, or it may be a decision not to explore, e_0. The result of exploration will be an observation which is a function of the decision e and the unknown state of

Figure 9.3. Geological technology 2. Mineral resource exploration.

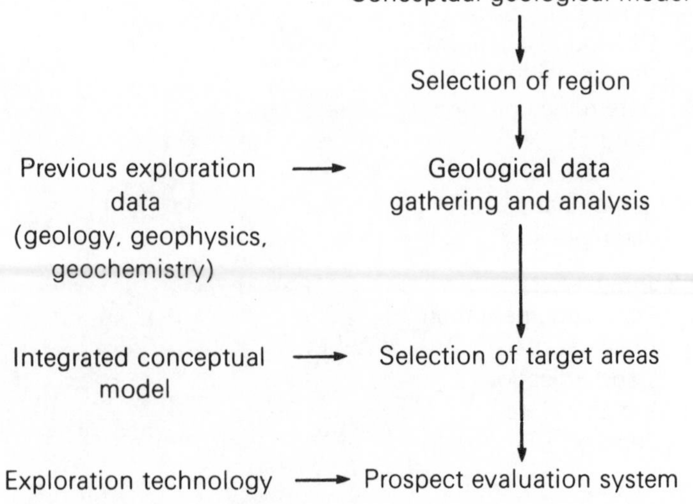

Mineral trade → Economic/political concepts
↓
Mineral/species type
↓
Conceptual geological model
↓
Selection of region
↓
Previous exploration data (geology, geophysics, geochemistry) → Geological data gathering and analysis
↓
Integrated conceptual model → Selection of target areas
↓
Exploration technology → Prospect evaluation system

nature θ. Similarly, the mining decision may consist of different options M_1, M_2 (e.g. large low-grade mine, selective operation) or M_0 (not opening a mine). Rendu sees the mining decision M being a function of the state of nature θ. This is convenient for the simplified decision tree, but in a real situation the state of the world (economically and politically) would be a prime consideration.

Consideration of the practical applications of the Engel Simulator shows that percentage profit is not the only consideration when deciding on an exploration programme. It is no use talking of huge profits if the exploration budget cannot meet the necessary investment cost. This issue has been taken up by Rendu (1976) in his explanation of 'utility functions'. The

Figure 9.4. Geological technology 3. Prospect evaluation system.

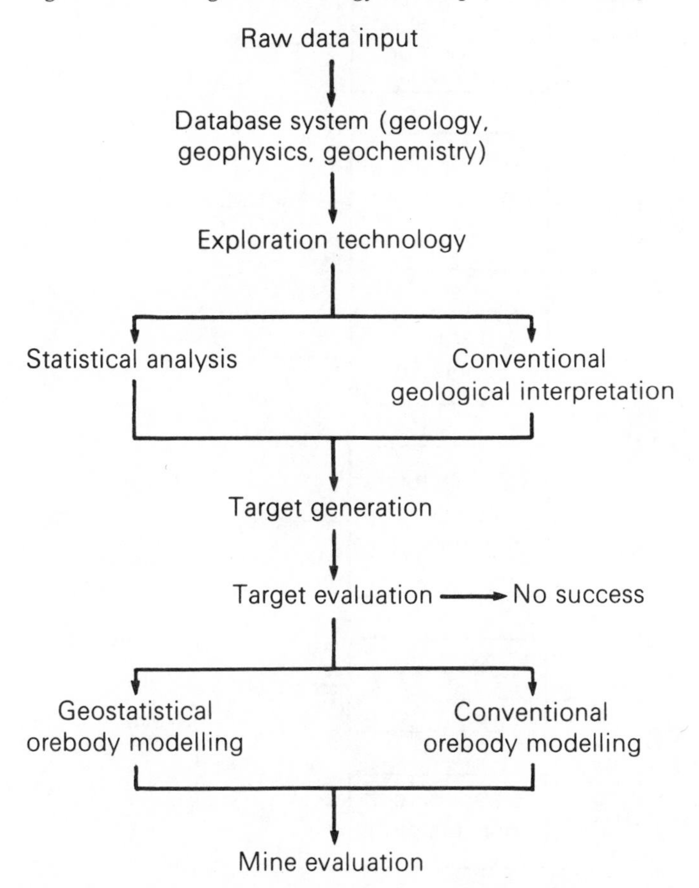

utility function represents the policy of an exploration company towards risk.

The 'decision trees' according to Koch & Link (1971) and Rendu (1976) are shown in Figures 9.5 and 9.6.

Slichter (1960) described prospecting as 'the world's biggest and best gambling business'. Even very large-scale exploration programmes cannot ensure success, and the concepts of 'gamblers' ruin' are familiar to all decision makers in exploration. The nature of this exploration gamble is such that the participants are encouraged to increase the odds of success. The geologist has therefore been encouraged to develop conceptual associations relating mineral deposits to geologic environments

Figure 9.5. Flow chart for decision making in the mineral industries. Adapted from Koch & Link (1971).

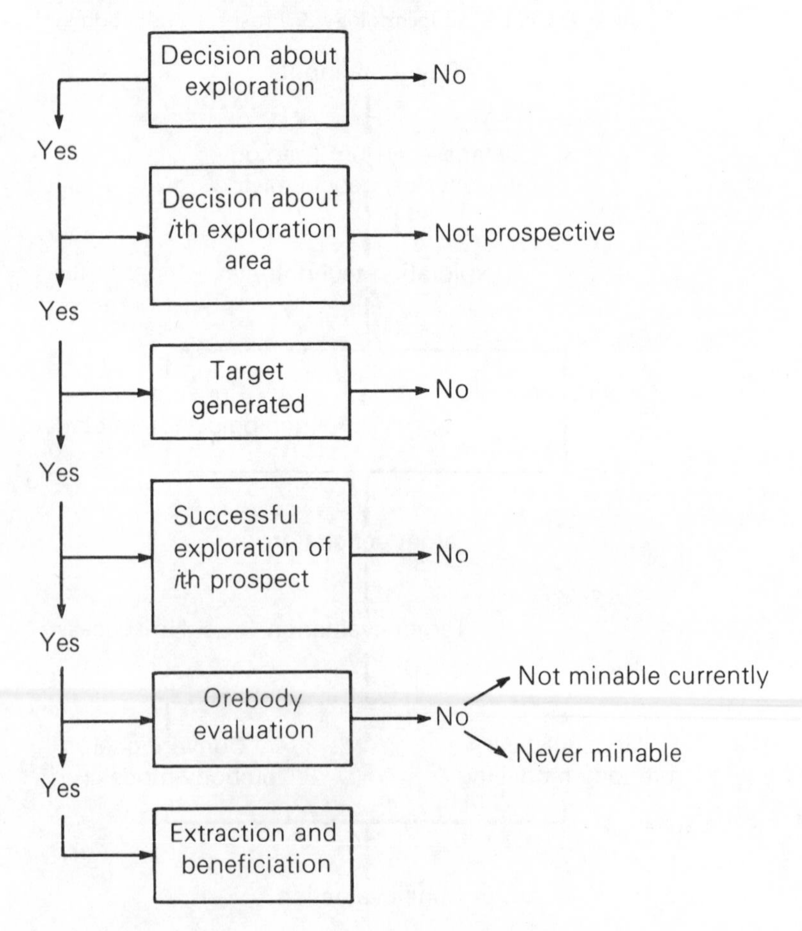

or specific time intervals. This has led to the development of many 'exploration maxims', which have given rise to many 'fashions' in exploration. Theories of genesis are often difficult to quantify and are not easily conformed to a pattern of exploration which has an emphasis on operations research and risk quantification.

Several factors have led to an interest in applying statistics in exploration, including advances in statistical theory, the general availability of computers to process large amounts of data, and a need for the quantification of exploration decisions. The correct usage of statistics is dependent on the availability of suitable raw data which is amenable to mathematical manipulation. Geological data are not always available in a numeric form and are often fragmentary being derived from limited surface outcrops, incomplete mapping or drill intersections.

Figure 9.6. Simplified decision tree. After Rendu (1976).

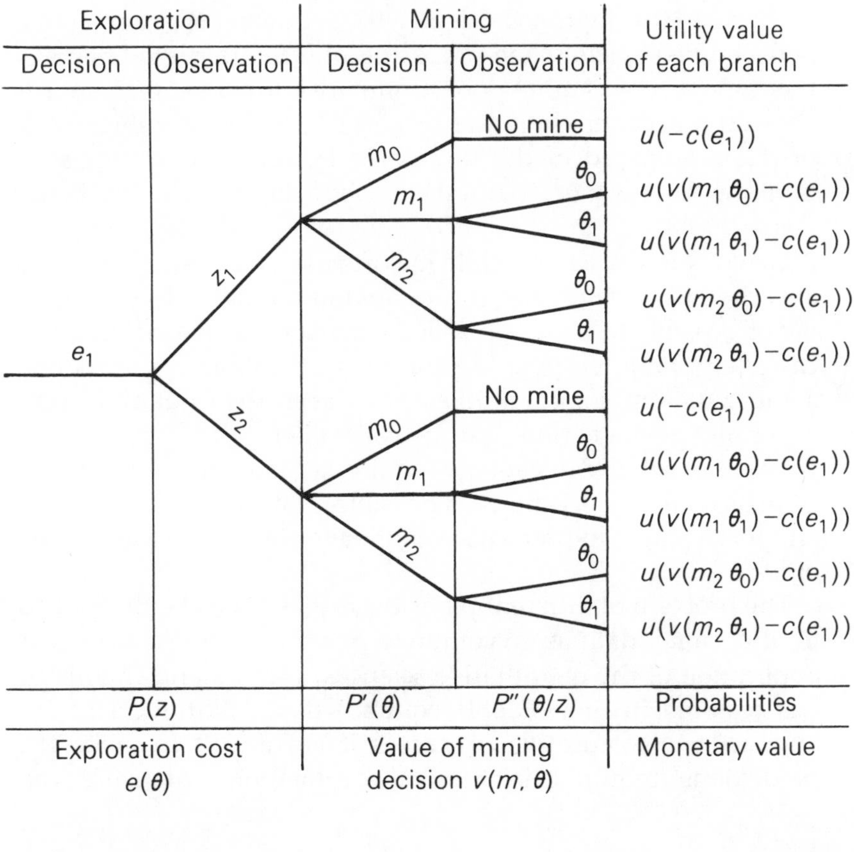

The added constraint of geologic time means that we cannot, for the most part, observe natural processes of ore emplacement due to the fact that:

(*a*) they are complete and finished;
(*b*) the processes are too slow to observe;
(*c*) surface evidence is removed, altered or obliterated;
(*d*) evidence at depth is inaccessible.

Geologic sampling is required to provide as much information as possible in an inexpensive manner, yet remain unbiased. Chadwick (1975*a*, *b*; 1976) showed that the tendency to see in information what one expects rather than what is really there afflicts geologists in particular, because they are used to drawing interpretations from limited data. He showed that the Poggendorf and Zollner illusions cause problems to geologists.

Another important factor contributing to the variability of geological data is that present in geologic maps prepared by different geologists, the maps being a product of the geologists' training, experience and bias. This point was well demonstrated by Burns, Huntington & Green (1977) showing that agreement between geologists when preparing photo-interpreted maps of the same area was quite poor. Comparisons between geologists showed a best correlation of some 12%, whereas comparisons of maps prepared of the same area by the same geologist in multiple runs – four in number – showed only slightly better results, 19%.

Geology as a science, while making use of the laws of physics and chemistry, differs from these two disciplines. In geology it is not possible to have controlled experiments in any but a few cases (e.g. high temperature and pressure laboratory work and some sedimentological studies), and even then not all factors, especially geologic time, can be controlled.

Problem solving in geology requires a theory to be developed from as many observations as possible. Not all observations will support a single hypothesis and not all future observations will either.

The problem of different plausible explanations being applied to the same data is a common one in geology. Even the application of the quantitative sciences, physics and chemistry, has resulted in unavoidable contradictions. Without a sound knowledge of causative processes it is difficult to make the predictions required in mineral exploration. 'Statistics can provide a means of analysing large amounts of data impartially and providing predictions within known limits of accuracy' Vistelius (1976).

The absolute definitions of acceptable financial risk and gain vary according to the dictates of the particular mining company. These parameters do define the type and size of ore deposit sought. This has naturally led to the development of concept-orientated exploration formats that entail the search for and application of 'significant' spatial relationships between geological features and ore deposits. Within this context of target generation, 'statistical' exploration is a powerful method of outlining 'favourable' areas. Various techniques have been developed that specifically relate to this type of exploration. Analytical techniques have been applied to both of the major areas of exploration activity, i.e. finding extensions to known mining areas/deposits, and the prospecting of virgin areas, or for virgin metals. In the former case, where there has been more than one period of production in one or more associated metals, the validity of the statistical analysis is easily verifiable. One or more of the early production periods can be analysed and exploration areas outlined; these can then be compared with the distribution of deposits discovered after the analysis period.

Any exploration model basically expresses two concepts, which can be simply formulated as (Harris 1967):

$$P(X) = F(X) \cdot G(X)$$

where X is a measure of the number of deposits (mines, dollar value etc.); $P(X)$ is the unconditional probability of the discovery of X deposits; $F(X)$ is the probability of occurrence of X deposits; and $G(X)$ is the probability of detection of X deposits. Both of the sub-models $F(X)$ and $G(X)$ can be regarded as independent.

The basis is to define a mathematical model to quantify the variables in mineral exploration and to optimise the profit objectives. This conceptually requires the combination of mineral occurrence and the effectiveness of the search. Many specific exploration models have been developed, often relying on control area concepts and prior grid spacing. In these models, the generation of probabilities is dependent upon the statistical base defined for a particular 'control area'. This control area therefore defines the meaningful variables and/or interaction of variables in the evaluation of the study area. The problem of defining a suitable control area is one that has received much attention. Even if one generates statistical bases upon several areas, each representative of geological 'type' or 'province', the problem is not alleviated. Figure 9.7 shows a conceptual flow chart for statistical methods of exploration.

Geostatistical crustal abundance models are frequency distribution functions of chemical elements in the Earth's crust. The average value represents the crustal abundance or 'clarke' of the particular element. Commonly only one element is considered at a time, although occasionally many elements may be considered simultanesously.

One of the great advantages of crustal abundance models is that they are relatively simple, although the basic assumptions on which they are based can only be approximately satisfied at best. However, McKelvey (1960) showed that a plot of reserves against crustal abundance for a wide variety of metals – plotted on a logarithmic scale – defined an approximate linear patterning, both for the USA and 'globally. Erickson (1973) utilised this model to define the potential recoverable reserves (R) of metal for the USA. His definition was:

$$R = 2.45 \times 10^6 C \text{ t}$$

(C = p.p.m. crustal abundance). Erickson used lead as a standard. Earlier, Ovchinnikov (1971) applied McKelvey's model on a global basis and defined the relationship:

$$R = 3.3 \times 10^6 C \text{ t}$$

His constant was defined by averaging results for 32 metals. If R in both Erickson's and Ovchinnikov's equations are divided by the total area of the region considered, giving a unit regional value (UR) a better comparison can be made:

$$\text{UR} = 0.022\ C \text{ t/km}^2 \text{ (Ovchinnikov)}$$
$$\text{UR} = 0.313\ C \text{ t/km}^2 \text{ (Erickson)}$$

It should be noted that both of these estimates are based on figures for reserves only. When past production figures are also included in the value for R the UR figures are considerably altered; a comparison is given in Table 9.1.

Garret (1978) defined resource estimates related to crustal abundances for ten metals in Canada; the relationships are shown in Figure 9.8. Also plotted in this figure are the relationships of Erickson and Ovchinnikov. Garret used five categories of resource estimate, each estimate being cumulative in that it includes those below it. The lowest estimate (1) is of production to 1975, the second (2) includes reserves, and (3) resources.

These estimates above generally indicate that McKelvey's

Figure 9.7. Conceptual flow chart for statistical methods of exploration. Adapted from Harris (1966).

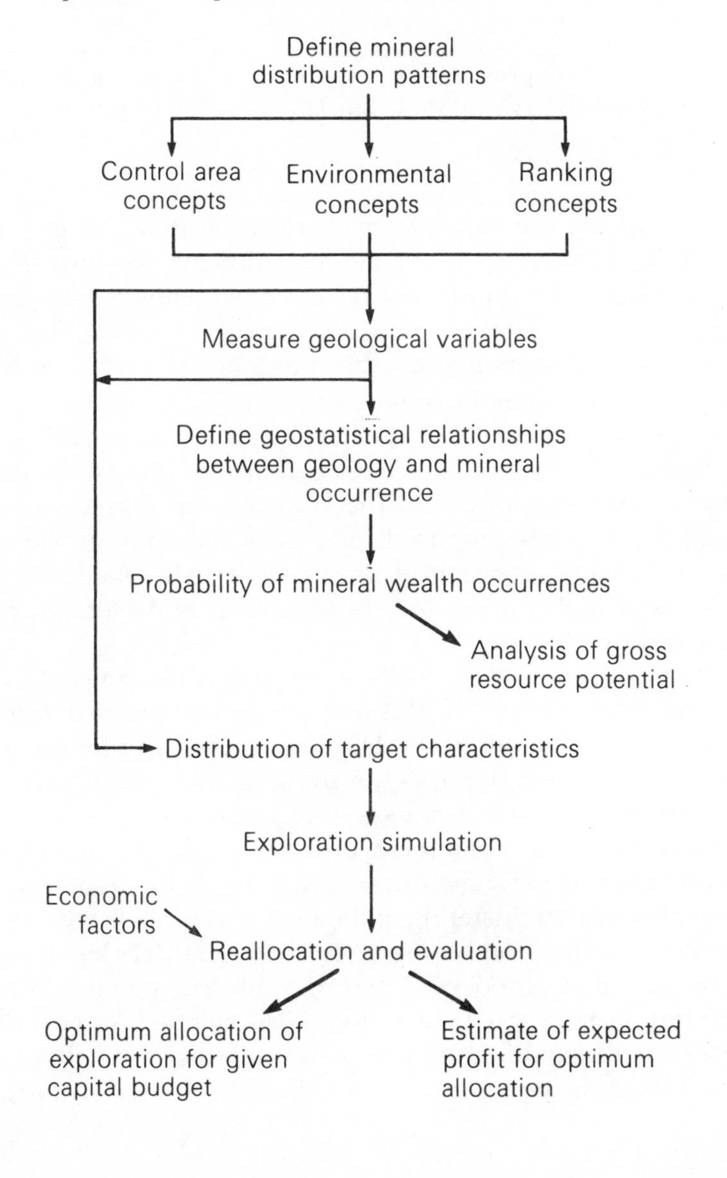

Table 9.1 *A comparison of unit regional values*
Values in tonnes/km^2

	Reserves only*		Reserves plus past production	
	McKelvey (1960) USA	Sekine (1962) Japan	Sekine (1962) Japan	Agterberg & Divi (1978) Canada
Copper	2.9	4.0	17.6	13.0
Lead	0.9	1.9	3.5	23.8
Zinc	2.5	8.9	53.8	50.8

*The figures are approximately proportional to the crustal abundance figures: respectively, 63 p.p.m. Cu, 12 p.p.m. Pb, 94 p.p.m. Zn (Lee & Yao 1965).

concept of a linear relationship between *R* and *C* is generally satisfied. However, it would be difficult to formalise a relationship of this kind since resource estimates are simply too nebulous.

Many recent workers have used lognormal models to relate metal concentration in ores to those in common rocks, e.g. Brinck (1971), Drew (1977), Deffeyes & MacGregor (1980). The weakness of this type of model is that it is difficult to define the reference crustal block, the control area. One of the first lognormal models proposed, and even now one of the most comprehensive, was that of Brinck (1971). He related the size of ore bodies, their frequency of occurrence and average grades for a variety of metals.

Of critical importance to these models is the concept of the unit regional value (UR). If it is assumed that mineral deposits are discrete events then the UR is meaningful in terms of the number of mineral deposits per unit area of constant size. The UR can be measured in a variety of ways and need not relate solely to the number of mineral deposits; monetary values could also be utilised, or some other parameter of the deposits. The concept of spatial clustering of deposits was formalised by Allais (1957) following from earlier work by Nolan (1936). Using a lognormal model for deposit size in combination with a Poisson distribution for spatial clustering Allais defined a generalised frequency distribution model for the amount of metal per unit area. Other clustering models that have been used include the exponential model of Slichter (1960) and the negative binomial distribution model proposed by Griffiths (1966) and Wilmott, Hazen & Wertz (1966), and which is now a commonly used model.

Figure 9.8. Resource estimates of ten metals in Canada. After Garrett (1978).

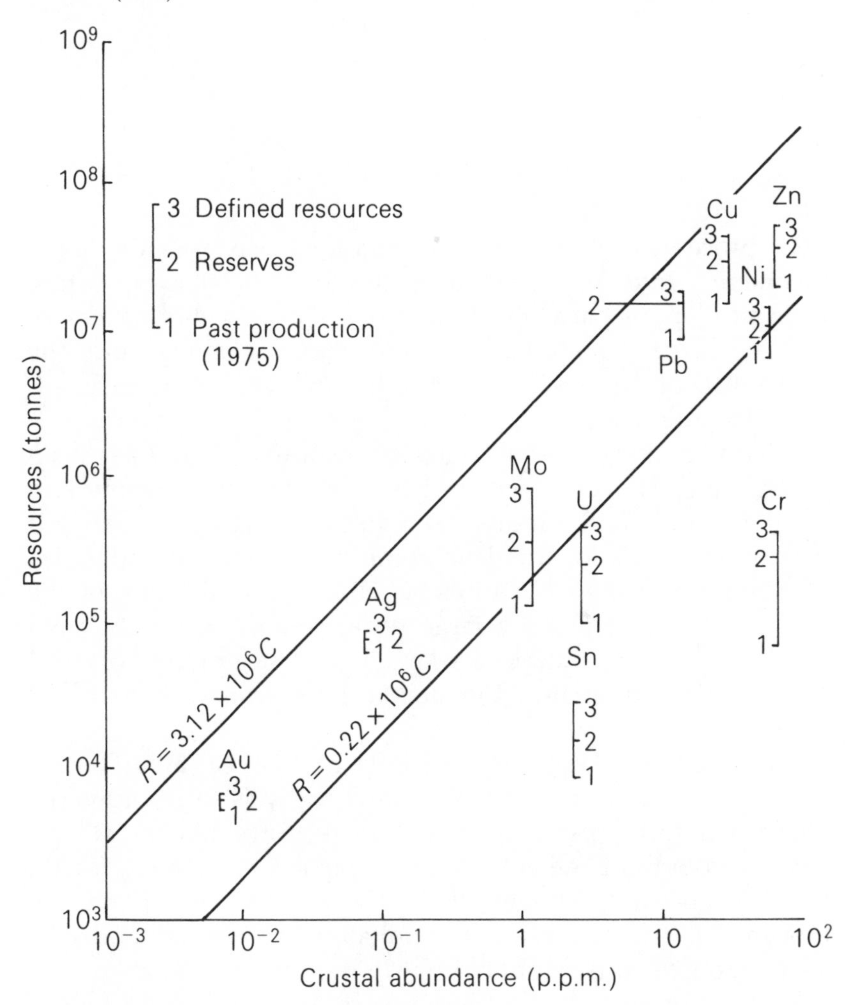

10

Some thought for the future

The processes of resource evaluation also include some incursions into the field of futurology to discern future requirements of particular resources. Not only it is necessary to discern the potential for resource availability and the probability of exploration success, but it is vital to try to assess the future demand.

In respect of mineral and energy resources there have been several major events which have affected production and demand. Initially the industrial revolution stimulated demand for raw materials for the developing manufacturing industries of Europe and then N. America. The industrialisation of the Western world required a large resource base to satisfy what then was an expanding market. The development of steel technologies stimulated the demand for iron ore, coal and limestone.

The First World War, the emergence of Russia as an industrial nation, the Second World War and the post-war industrial recovery in Europe and Japan kept the resource demand at high levels throughout the 20th century. Japan's economic growth, the Korean and Vietnam wars, the national independence movements and processes of nationalisation have all tended to keep resource consumption at fairly high levels.

Table 7.1 outlines the changing world metals demand for the period 1900–80. Aluminium is the 'wonder metal' with a demand growth rate of 11.4% between 1900 and 1950 and 7.9% between 1950 and 1980. These patterns of growth have provided the stimulus for the ideas of continuing growth as an indication of a 'healthy' economy. Hidden within these figures, however, is the fact that the consumption patterns for these major metals have shifted and will continue to do so for the next 20 or 30 years. In 1950 the percentage shares of metal consumption for the industrialised (Western) nations, the lesser developed (third world) countries and the Eastern Bloc were 82%, 4% and 13%, respectively. By 1980 the pattern had shifted to respective percentage shares of 61%, 12% and 27%. This represents a significant shift in resource consumption pattern. This pattern is also reflected in steel usage figures. Between 1970 and 1980 the steel usage *per capita* in the USA declined 16.7%, in terms of per dollar of GNP the decline was 32.3%. The corresponding figures for Western Europe are 17.2% and 33.8%, SRI (1985).

What this represents is the fact that there has been and will continue to be a change in the economics of base metal production. The shift in usage from industrialised to developing nations and the Eastern Bloc means essentially that there is in progress a paradigmal shift in resource economics, from what might be termed 'Western' economic principles to state enterprise principles. Added to this is also the factor, discussed earlier, of the changing patterns and pathways to the industrialised state. The pathway to industrialisation is not as energy intensive or as metal intensive as the pathway followed by the USA or perhaps Western Europe. Put another way, the same can be done with less.

This has been expressed by two analyses in this format: 'We don't anticipate any more pounds of metal being used in the world in 1990 than in 1980' (Carnegie, 1984); and 'We may be witnessing an historical change and the first major impact of the shift from an energy economy to an information economy . . . the implications would be startling. It would render absurd all the forecast of the past 20 years in respect to demand for energy, raw materials and food' (Drucker, 1985). An information society is a far less resource intensive society than the more 'traditional' industrial society.

As described earlier the traditional approach to forecasting has been to define explicit relationships between resource consumption and macro-economic parameters expressing the former as some function against time. The aggregate of specific end-use sectors are then commonly utilised to corroborate the macro-economic indicators such as consumption *per capita*.

If, however, as Drucker states, there is a decoupling from the traditional perspective (and the evidence seems to indicate this), then the traditional economic indicators are of no value to demand forecasting. This tends to indicate that there is no fundamental demand for any particular resource. There are functional demands, e.g. cooking utensils, which are the key drivers in resource demands. Forecast demands should therefore be made in terms of which materials are better suited to particular functions. This then allows for a greater role and for the evaluation of substitutes. Another way of putting it is that new technologies create new materials which perform particular functions better than traditional resources. Fibre optics and high temperature ceramics are particular examples of new technologies that will impact greatly on the metal demand scenarios of the future.

The worldwide economic restructuring that was a consequence of the 1973 oil crisis will continue as the trend is toward 'high technology materials'. As a result there could well be fundamental changes in the energy industries as less energy intensive materials are utilised even though they may have a petrochemical base.

The Stanford Research Institute, International, at a seminar given in Melbourne March 1985 suggested three technologies that could affect our society as much or greater than the microchip. These are, rapid solidification technology, high temperature ceramics and composite materials. Rapid solidification technology has the capacity to produce ductile metal alloys with very high tensile strengths at high temperatures. Another variant are amorphous metals which potentially are the strongest and most easily magnetised materials known. High temperature ceramics will drastically alter the automobile industry in the near future, and ceramic machine cutting tools should also soon be available. Ceramic coated materials designed to increase corrosion resistance and attrition loss should also soon be commercially available. Composite materials are now utilised in the design and building of aircraft.

Another major factor in the resource evaluation situation is the current state of the Western trading and merchant banks. Toward the end of the 1960s the liquid savings of industry was tending toward depletion and the shortage of liquidity was starting to show in the banking system. For example, the ratio of liquid assets to domestic assets for many national banks started to decline. During the early 1970s the real world money supply situation showed signs of very great instability. Concomitant with this was the fact that unemployment in the industrial nations started to rise from about 3 to 4% to about 8% (average), SRI (1985).

To counteract the trend to increasing unemployment many 'industrial' governments elected to suppress interest rates, which led to increased deficit spending and encouraged low interest loans to the lesser developed nations. Naturally this further depleted the liquidity of the Western banks and decreased the money supply quite drastically. By the early 1980s the banking system was close to collapse and the banks promptly increased interest rates to avert a collapse.

The increase in interest rates depressed economic activity and curtailed many energy and mineral projects. It also lowered the demand for metals. The current legacy therefore is an industrialised 'Western' world with low levels of liquidity, enormous trading deficits and international debts, very acute trade imbalances and the problems that a very powerful US dollar tends to develop. The outlook for metals trade is bleak.

There have thus been drastic and quite irreversible changes in the metal and energy industries. For many commodities the lesser developed countries have assumed the more dominant role both in terms of raw material and process product supplies. The dynamic markets are therefore going to be in these countries. In addition, tailored technologies to functional demands will significantly alter mining and processing resource requirements. The energy and mineral resource systems evaluations of the future will therefore be according to new non-traditional stratagems and economic criteria.

Appendix

Units and energy equivalents

Data Sources: Finney (1976), Dept. National Development and Energy, Canberra 1980, 1981.

Units

J = joule = unit of energy: the work done when a force of one newton (N) is displaced through a distance of one metre; a newton is the force required to give a mass of one kilogram an acceleration of one metre per second per second.

Table A.1 *Prefix abbreviations*

10^{18}	exa (E)	10^{3}	kilo (k)
10^{15}	peta (P)	10^{0}	
10^{12}	tera (T)	10^{-3}	milli (m)
10^{9}	giga (G)	10^{-6}	micro (μ)
10^{6}	mega (M)	10^{-9}	nano (n)

Table A.2 *Conversion factors*

Energy

1 J = 2.778×10^{-7} kWh = 9.478×10^{-4} BTU = 0.239 cal
1 kWh = 3.6×10^{6} J = 3.412×10^{3} BTU = 8.598×10^{5} cal
1 BTU = 1.055×10^{3}J = 2.931×10^{-4} kWh = 252.0 cal
1 cal = 4.187 J
1 erg = 10^{-7}J
1 watt year = 3.154×10^{7}J

Energy density (quantity per unit area)
1 MJ/cm^2 = 278 kW/cm^2 = 23.9 cal/cm^2
1 kJ/m^2 = 2.39×10^{-6} cal/m^2 = 10^{6} erg/cm^2

$1\ kWh/m^2 = 3.6\ MJ/m^2$
$1\ cal/cm^2 = 1\ langley = 10\ k\ cal/m^2 = 0.0419\ MJ/m^2 = 11.6\ Wh/m^2$
$= 3.69\ BTU/ft^2$
$1\ BTU/ft^2 = 1.1355\ J/cm^2 = 11.355\ MJ/m^2 = 0.271\ cal/cm^2$
$= 3.14\ Wh/m^2$

Power (energy per unit time)
$1\ W = 1\ J/s = 10^7\ erg/s = 0.239\ cal/s$
$1\ erg/s = 10^{-1}\mu\ W = 1.43 \times 10^{-6}\ cal/min$
$1\ cal/min = 0.0698\ W = 6.98 \times 10^4\ \mu W = 6.98 \times 10^5\ erg/s$
$1\ cal/s = 4.187\ J/s$

Energy intensity (power per unit area)
$1\ W/m^2 = 3.6\ kJ/m^2/h = 9.26 \times 10^{-5}\ kWh/ft^2/h = 0.316\ BTU/ft^2/h$
$= 0.086\ cal/cm^2/h = 10^3\ erg/cm^2/s$
$1\ W/cm^2 = 10^7\ erg/cm^2/s$
$1\ \mu W/cm^2 = 10\ erg/cm^2/s = 1.43 \times 10^{-5}\ cal/cm^2/min$
$1\ kW/cm^2 = 3.6\ kJ/cm^2 = 0.0036\ MJ/m^2 = 0.086\ cal/cm^2$

Q (the total expected world consumption of energy in the year 2000)
$Q = 10^{18}\ BTU = 10^{21}\ J = 3 \times 10^{14}\ kWh = 3 \times 10^{10}$ t of black coal
$= 2 \times 10^{11}$ barrels of oil

Energy Equivalents of Fuels

Oil

- 1 barrel $= 6.3 \times 10^9\ J = 5.7 \times 10^6\ BTU$
- 1 tonne $= 4.47 \times 10^{10} J$
- 1 Mtoe (million tonnes of oil equivalent)
  $= 4.47 \times 10^{16} J = 1.24 \times 10^{10} kWh = 4.24 \times 10^{13} BTU$

Gasoline

- 1 l $= 3.5 \times 10^7\ J$

Natural Gas

- 1 $ft^3 = 10^3\ BTU = 1.05 \times 10^6\ J$
- 1 $m^3 = 3.7 \times 10^7\ J$

Black Coal

- 1 ton $= 2.8 \times 10^{10}\ J = 2.6 \times 10^7\ BTU$
- 1 kg $= 3.1 \times 10^7\ J$
- 1 lb $= 1.3 \times 10^4\ BTU$
- 1 Mtce (million tonnes of black coal equivalent)
  $= 2.85 \times 10^{16} J = 7.92 \times 10^9 kWh = 2.75 \times 10^{13} BTU$

Uranium metal

1 ton $= 8.1 \times 10^{16}$ J (fully burnt)

References

Agterberg, F. and Divi, S. 1978. *Econ. Geol.* 73, 230–45

Allais, M. 1957. *Management Science* 3, 285–347.

Arnoux, L. 1979. In *Goals and Guidelines*, Proceedings of Seminar on Energy: Environmental Perspectives. MOE Wellington, New Zealand, pp.303–18

Arnoux, L. 1981. *Energy, Decision-Making and Equity.* Presented to ANZAAS Conference, May 11–15, Brisbane, Australia

Austin, J. 1974. *Can. Inst. Min. Metall. Bull.* 197, 49–56

Barnett, D. 1979. *Minerals and Energy in Australia*. Cassell, Australia

Barnett, R. 1980. *National Times, Australia*, nos 489–92, June/July

Becker, P. 1980. *Aust. Inst. Energy 2nd Annual Conf.*, May, Melbourne, Australia, pp.7–16

Beckerman, W. 1977. In *Solar Australia: Australia at the Crossroads* Ambassador Press, Australia, pp.116–21

Berg, C. 1974. *Technol. Rev.* Feb., 15–23

Berry, R. and Fels, M. 1973. *Bull. Atom. Sci.* Dec. 11–17, 58–60

Brain, P. and Schuyers, G. 1981. *Energy and the Australian Economy.* Cheshire, Melbourne

Brinck, J. 1971. *Eurospectra* 10, 45–6

Bunyard, P. 1976. *The Ecologist* 6:3, 87–100

Burns, K., Huntingdon, J. and Green, A. 1977. *APCOM '77 15th International Symposium*, pp.275–86

Capra, F. 1982. *The Turning Point*. Wildwood House, London

Carnegie, Sir R. 1984. *Business Rev. Weekly* Nov. 5–9

Chadwick, P. K. 1975*a*. *Nature*, 256, 570–3

Chadwick, P. K. 1975*b*. *New Sci.* 18/25, 728–31

Chadwick, P. K. 1976. *Nature* 260, 397–401

Collingridge, D. 1980. *The Social Control of Technology.* St Martin's Press, New York

Collier, J. 1983. *Pacific Basin Econ. Conf.*, Santiago, Chile. (Supplementary paper)

Commoner, B. 1978. *Energy and Labour: Job Implications of Energy Development or Shortage*. Address: The Canadian Labour Congress, Feb. 20, Ottawa

Cook, E. 1971. The flow of energy in an industrial society. In *Energy and Power*. Scientific American pp.83–91.
CURA (Centre for Urban Research and Action). 1982. *Social Impact Studies Series* 1
Deffeyes, K. and MacGregor, I. 1980. *Sci. Am.* 242 (1), 66–76
Dick, J. and Mardon, C. 1979. In *Energy and People* (Diesendorf, M., ed.). Society for Social Responsibility in Science, Canberra, pp.7–18
Diesendorf, M. (ed.). 1979. *Energy and People*. Society for Social Responsibility in Science, Canberra
Drew, M. 1977. *Resources Policy* March, 60–6
Drucker, P. 1985. *Wall St. J.* Jan.
Erickson, R. 1973. *US Geol. Survey Prof. Paper* 820, 21–5
Etheridge, W. 1979. *Aust. I M M Annual Conf.*, Perth, Australia, pp.269–90
Evans, D. and Atkins, A. 1979. In *Energy and People* (Diesendorf, M., ed.). Society for Social Responsibility in Science, Canberra, pp.141–6
Finney, A. 1976. *University of Tasmania Environmental Studies* 2
Garrett, R. 1978. *J. Math. Geol.* 10 (5), 481–94
Grathwohl, M. 1982. *World Energy Supply*. W. de Gruyter, Berlin
Govett, M. and Govett, G. 1976. *World mineral supplies: Assessment and Perspectives*. Elsevier, Amsterdam
Griffiths, J. 1966. *Oper. Res.* 14, 189–209
Harris, D. 1966. *Trans. Soc. Min. Eng.* 199–216
Hawken, P. 1984. *The Next Economy*. Angus & Robertson, Sydney
Haynes, D. 1979. In *Mineral Resources of Australia* (Kelsall, D. and Woodcock, J., eds). Australian Academy of Technological Sciences, Melbourne, pp. 73–96
Higgins, R. 1978. *The Seventh Enemy*. Pan Books, London.
Hoyle, F. 1977. *The Ten Faces of the Universe*. W.H. Freeman, San Francisco.
Illich, I. 1974. *Tools for Conviviality*. Caldar and Boyars, London
Illich, I. 1973. *Energy and Equity*. Caldar and Boyars, London
Koch, G. and Link, R. 1971. *Statistical Analysis of Geological Data* (three volumes). J. Wiley & Sons, New York
Lee, T. and Yao, C. 1965. *Int. Geol. Rev.* 12 (7), 778–86
Lovins, A. 1977. *Soft Energy Paths*. Pelican Books
Mackrell, K. 1983. *Proc. Aust. Inst. Eng. Conf. Energy '83*, pp.79–84
Mardon, C., White, D., Sutton, P., Pears, A. Dick, J. and Crow, M. 1978. *Seeds for Change*. Patchwork Press, CCV, Melbourne.
McKelvey, V. 1960. *Am. J. Sci.* 258-A, 234–41
MacKenzie, D. 1981. *Aust. IMM Ann. Conf.*, Syndey, Australia, pp.85–99
McLennan, Sir I. 1978. In *International Resource Management* (Woodcock, J., ed.). Aust. IMM, pp.11–17
Messel, H. (ed.) 1979. *Energy for Survival*. Pergamon Press, N.S.W.
Miller, J. 1981. *Aust. Inst. Energy Ann. Conf.*, Sydney, Australia, Appendix A
Musgrove, A., Stocks, K., Essam, P., Le, D. and Hoetzl, J. 1983. *CSIRO Division of Energy Technology Technical Report* 2
Nolan, T. 1936. *US Geol. Survey Bull.* 871. 5–77
Ophuls, W. 1977. *Ecology and the Politics of Scarcity*. W.H. Freeman, San Francisco.

O'Riordan, T. 1976. *Environmentalism*. Pion Ltd.
Ovchinnikov, L. 1971. *Dokl. Acad. Nauk. SSSR* 196 (1–6, 200–3)
Park, C. Jr. 1975. *Earthbound*. Freeman, Cooper & Co., California
Pearson, D. 1977. (Quoted by West, J. and Tapp, B. 1979. *Aust. Mining* 71 (9), 51–62)
Rawlings, B. (Convenor) 1983. *Inst. Eng. Aust.* Pub. 83/2, pp, 53–77.
Rendu, J. 1976. In *Advanced Geostatistics in the Mining Industry* (Guarascio, M., David, M. and Huijbregts, C., eds). D Reidel, Dordrecht
Rens, I. 1982. *Ecology and Politics*. The Geelong Lectures, Deakin University, Victoria, Australia.
Roby, K. 1979. In *Prospect 2000* (Waddell, S., ed.) ANZAAS Conf., May, Perth, pp. 121–35.
Rutland, R. 1987. *Australian Mining* 39, 41.
Saunders, D. 1976. *Fossil Fuels*. (Quoted in *Aust. Inst. Energy Proc. Pub.* 77/6, figure 1, p. 3). State Elec, Commision, WA
Schumacher, E. 1974. *Small Is Beautiful*. Abacus Books, Australia
Sekine, Y. 1962. *Mining Geol. (Jpn.)* 12 (56), 16–26
Shiels, G. 1982. *Energy Paper 2. Aust. Inst. Energy Special Publ.*, February
Slichter, L. 1960. *Min. Eng.* 12, 570–6
SRI (Stanford Research Institute International). 1985. *The Global Mineral Industry* (seminar), March, Melbourne
Stewart, E. (Convenor). 1983. *Aust. Inst. Eng. Energy '83 Conf.*, pp.18–43
Temple, D. 1979. *Bull. Aust. Inst. Min. Metall.* 436/7, 9–13
Vistelius, A. 1976. *J. Geol.* 84, 629–51
Watt, K. 1979. In *Prospect 200* (Waddell, S., ed.). *ANZAAS Conf.*, May, Perth, pp.200–24
West, J. 1978. MSc thesis. James Cook University of North Queensland
White, D. 1979. *Aust. IMM Ann. Conf.*, Perth, pp.339–48
Wilmott, R., Hazen, S. Jr. and Wertz, J. 1966. *6th APCOM Symposium*, Pennyslvania State University

Recommended further reading

There are a number of books that we would recommend for additional reading to develop further the scope of the concepts and issues raised.

For the interested reader wishing to develop a study in systems analysis and methodology, we would suggest Bennett & Chorley as a very suitable beginning. The book by Häfele & Kirchmayer follows on from this in the sense of providing an introduction to the modelling of large-scale energy systems.

An overview of energy in terms of the resource, technology and economic perspectives is provided by Grathwohl. A more embracing overview of resources and the implications resultant from developing them is given by Simon & Kahn. The books by Anderson and Hawdon provide a political and economic view of the energy industry in both sectoral and global perspectives.

The books by Birrell *et al.*, Brain & Schuyers and Saddler cover many of the above issues but relate specifically to the Australian context. Hawken and Capra provide an economic and sociological commentary on the changes current within western societies. As such they provide a backdrop for energy and mineral resource system analyses. Some of the future trends discussed in these two books, if brought to fruition, would significantly alter the energy and mineral systems, hence economic stability of many nations.

It should be noted that this reading list is by no means exhaustive or comprehensive but is merely a guide to further study.

Anderson, J. 1984. *Oil*. Sidgwick & Jackson, London
Bennett, R. and Chorley, R. 1978. *Environmental Systems*. Methuen & Co., Ltd.
Birrell, R. Hill, D. and Stanley, J. (eds). 1982. *Quarry Australia*. Oxford University Press, Melbourne.
Brain, P. and Schuyers, G. 1981. *Energy and the Australian Economy*. Longman Cheshire, Melbourne
Capra, F. 1982. *The Turning Point*. Wildwood House, London
Grathwohl, M. 1982. *World Energy Supply*. W. de Gruyter, Berlin
Häfele, W. and Kirchmayer, L. 1980. *Modelling of Large Scale Energy Systems*. Pergamon Press.
Hawdon, D. (ed.). 1984. *The Energy Crisis: ten years after*. Croom Helm, London
Hawken, P. 1984. *The Next Economy*. Angus & Robertson, Sydney
Saddler, H. 1981. *Energy in Australia*. Allen & Unwin, Sydney
Simon, J. and Kahn, H. 1984. *The Resourceful Earth: A Response to Global 2000*. Blackwell, Oxford
Worldwatch Institute. *The State of the World*.

Papers and articles of interest are to be found in the following journals: *Resources and Energy; Energy Policy; Fuel; Energy World; Journal, Institute of Fuel; Journal, Institute of Energy.*
An occasional article of popular interest can be found in: *American Scientist; New Scientist; Discovery; Science; Nature.*

Index